T0200004

Peirce Mattering

American Philosophy Series

Series Editor: John J. Kaag, University of Massachusetts Lowell

Advisory Board

Charlene Haddock Siegfried, Marilyn Fischer, Scott Pratt, Douglas Anderson, Erin McKenna, and Mark Johnson

The *American Philosophy Series* at Lexington Books features cutting-edge scholarship in the burgeoning field of American philosophy. Some of the volumes in this series are historically oriented and seek to reframe the American canon's primary figures: James, Peirce, Dewey, and DuBois, among others. But the intellectual history done in this series also aims to reclaim and discover figures (particularly women and minorities) who worked on the outskirts of the American philosophical tradition. Other volumes in this series address contemporary issues—cultural, political, psychological, educational—using the resources of classical American pragmatism and neo-pragmatism. Still others engage in the most current conceptual debates in philosophy, explaining how American philosophy can still make meaningful interventions in contemporary epistemology, metaphysics, and ethical theory.

Recent Titles in the Series

Peirce Mattering: Value, Realism, and the Pragmatic Maxim, by Dorothea Sophia

Humanism, Antitheodicism, and the Critique of Meaning in Pragmatist Philosophy of Religion, by Sami Pihlström

A Pragmatist Philosophy of History, by Marnie Binder

Quine, Conceptual Pragmatism and the Analytic-Synthetic Distinction, by Robert Sinclair

William James's Essays in Radical Empiricism: *A Critical Edition*, by H.G Callaway

Dewey and the Aesthetic Unconscious: The Vital Depths of Experience, by Bethany Henning

Pragmatic Philosophy of Religion: Melioristic Case Studies, by Ulf Zackariasson

Endurance Sport and the American Philosophical Tradition, by Douglas R. Hochstetler

Pragmatist and American Philosophical Perspectives on Resilience edited by Kelly A. Parker and Heather E. Keith

Reconstructing the Personal Library of William James: Markings and Marginalia from the Harvard Library Collection by Ermine L. Algaier IV

Rorty, Religion, and Metaphysics by John Owens

Ontology after Philosophical Psychology: The Continuity of Consciousness in William James's Philosophy of Mind, by Michela Bella

Richard Rorty and the Problem of Postmodern Experience: A Reconstruction, by Tobias Timm

The Pragmatism and Prejudice of Oliver Wendell Holmes, Jr., edited by Seth Vannatta

Peirce and Religion: Knowledge, Transformation, and the Reality of God, by Roger A. Ward

William James, Moral Philosophy, and the Ethical Life, edited by Jacob L. Goodson

Epistemic Issues in Pragmatic Perspective, by Nicholas Rescher

Loving Immigrants in America: An Experiential Philosophy of Personal Interaction, by Daniel G. Campos

Peirce Mattering

Value, Realism, and the Pragmatic Maxim

Dorothea Sophia

LEXINGTON BOOKS
Lanham • Boulder • New York • London

Published by Lexington Books
An imprint of The Rowman & Littlefield Publishing Group, Inc.
4501 Forbes Boulevard, Suite 200, Lanham, Maryland 20706
www.rowman.com

86-90 Paul Street, London EC2A 4NE, United Kingdom

British Library Cataloguing in Publication Information Available

Library of Congress Cataloging-in-Publication Data Available

ISBN 9781793654106 (cloth : alk. paper) | ISBN 9781793654113 (epub)

To Bonnie
whose egg I made in the process of making my daughter, Sybil,
many years ago.

Contents

List of Figures

Abbreviations

CLL   *Chance, Love, and Logic: Philosophical Essays*, edited by Morris R. Cohen (1923), NY: Harcourt Brace, followed by page number.

CP   *Collected Papers of Charles Sanders Peirce*, edited by Charles Hartshorne and Paul Weiss (1931–1935) (volumes 1–6) and Arthur Burks (1958) (volumes 7–8), Cambridge, MA: Harvard University Press, followed by volume and paragraph number.

EP   *The Essential Peirce: Selected Philosophical Writings*, edited by Nathan Houser et al. (1992 & 1998) (volumes 1–2), Bloomington, IN: Indiana University Press, followed by volume and page number.

MS   *The Charles S. Peirce Papers* in Houghton Library (1966) followed by a Richard Robin number. Manuscript letters and drafts are identified by an L preceding the manuscript number.

N   *Contributions to "The Nation" Volumes 1–4,* edited by Kenneth Laine Ketner and James Cook (1975–1987) Lubbock, TX: Texas Technological University Press.

PAP   *Pragmatism as a Principle and Method of Right Thinking: The 1903 Harvard Lectures on Pragmatism,* edited by Patricia Ann Turrisi (1997) NY: State University of NY.

PM   *Philosophy of Mathematics: Selected Writings Charles S. Peirce*, edited by Mathew E. Moore (2010a) Bloomington, IN: Indiana University Press, followed by page number.

PWP   *Philosophical Writings of Peirce,* edited by Justus
Buchler (1940, 1955) NY: Dover.

RLT   *Reasoning and the Logic of Things: The Cambridge
Conferences Lectures of 1898,* edited by Kenneth Laine
Ketner (1992) Cambridge, MA: Harvard University
Press, followed by page number.

SS   *Semiotic and Significs: The Correspondence between
Charles S. Peirce and Victoria Lady Welby,* edited by
Charles S. Harwick (1976) Bloomington, IN: Indiana
University Press.

SW   *Charles S. Peirce Selected Writings: Values in a
Universe of Chance,* edited by Philip P. Wiener (1958)
NY: Dover, followed by page number.

W   *Writings of Charles S. Peirce: A Chronological Edition*
(volumes 1–6 & 8) edited by the Peirce Edition Project
(1982–2010), Bloomington, IN: Indiana University
Press, followed by volume and page number.

In referring to these, I observe the conventional notation set forth for referencing the primary texts of Peirce. The exception to this is Matthew E. Moore's *Philosophy of Mathematics: Selected Writings Charles S. Peirce,* which does not as yet have a convention; I have chosen PM to reference this text. In addition, whenever possible, I will add dates.

Introduction

What follows matters and it is "mattering."

What does it mean to say something, anything, everything matters and what is *it* that matters—what *is* "mattering"?

I first wondered about the meaning of "mattering" on coming across a small piece in a morning newspaper. When Australian lawyer and High Court judge Mary Gaudron, who in the 1970s was arguing for a female minimum wage, was asked by a fellow lawyer, "Does that matter?" she responded: "The matters that matter may be different, depending on who is doing the mattering" (SMH: 9 Dec 1986). It was many years later, trawling records of thinkers over the past three millennia that I came across one whose method of inquiry could meaningfully respond to such a question. This was Charles Sanders Peirce (1839–1914). Having a working knowledge of his systematic method of inquiry is crucial to understanding "mattering" so let me provide some detail here.

For those who are not familiar with Peirce, I have included a lengthy appendix outlining "Peirce's Architectonic Classes" of heuretic science and a second appendix: "The Primary Texts of Charles Sanders Peirce." An invaluable resource is available of some of Peirce's terminology in his own words: *The Commens Dictionary of Peirce's Terms* edited by Mats Bergman and Sami Paavola (2014) http://www.commens.org/dictionary. I strongly recommend you "bookmark" it for reference as you read.

Early in his philosophical writing, he responded to what *it is* in his paper "On a New List of Categories" (EP1.1-10, 1867)

based upon the theory already established [by Kant], that the function of conceptions is to reduce the manifold of sensuous impressions to unity and that the validity of a conception consists in the impossibility of reducing the content of consciousness to unity without the introduction of it. (EP1.1, 1867)

He explained in this paper and in a later rewrite (MS.403, 1893) that "IT in general is the conception of 'suchness' or present in general. In one of its meanings, Aristotle called this IT 'substance.' The junction at which IT as 'suchness,' joins 'thusness' of IS, is the concept called 'being.'. . . Thus," he says, "substance and being are the beginning and end of all conception." He pointed out, however, that

> substance is inapplicable to a predicate and being is equally so to a subject. . . .
> The conception of being arises upon the formation of a proposition. A proposi-
> tion always has, besides a term to express the substance, another to express the
> quality of that substance; and the function of the conception of being is to unite
> the quality to the substance. (EP1. 2, 4, 1867)

He identified subjects and predicates in his threefold intermediate categories between the manifold of impressions of "thusness" of *being* and the unity of the proposition of "suchness" of *substance*, as

Quality which is reference to a ground,
Relation which is reference to a correlate.
Representation which is reference to an Interpretant. (See EP1 6, 1867)

These three categories are the irreducible elements "as modes or tones of thought" (EP2. 247, c.1890) present in every phenomenon and are what Peirce came to call the Firstness, Secondness, and Thirdness of objective being. They reveal nascent meaning and can reveal "mattering" through extension of Peirce's method of inquiry, where inquiry is the activity engaged in to seek answers to genuine questions.

All inquiry relies on observation, reasoning, and experimentation. Peirce called inquiry which systematically seeks to discover whatever there may be that is true, science, and classified it into three branches:

1. Positive science—the science of things. This is experimental science: it is conducted for the purpose of determining the ontological validity of a hypothesis.
2. Semeiotic—the science of representations. This is the process of reasoning: it is conducted for the purpose of determining meaning. It entails creation, explication, and verification of a hypothesis.
3. Formal science—the science of forms. This is theoretical science: it is conducted for the purpose of identifying the means and the method for conducting inquiry.

In his *Teleological Logic* (MS.108, begun in 1865; W1: 303) in which he first classified science, Peirce began by defining logic. Let me paraphrase, amplify, and annotate his definition.

Logic, like science in general, is an activity—it is a process. In its narrow sense, logic is the process of attaining truth by way of identifying its necessary conditions. Peirce broadened this process to include the use of his system of signs as a means of identifying the *general* conditions producing meaning. Signs, as he defined them, are embodied cognition, that is, they are objective. He called his broadened logic *objective symbolistic*, or the semeiotic of symbols or simply semeiotic. Symbols are of three classes: *terms*, which call attention to things or quasi-things; *propositions*, which declare facts; and *arguments*, which profess to enlighten us as to the rational connections of facts or possible facts (MS[R].142.6, 1899-1900). They are regularities or laws, including power laws, of the indefinite future.

"Form is that respect in which a representation stands for a thing *prescinded* from all that can serve as the basis of a representation and therefore from its connection with the thing" (MS.108, 1865; W1: 303). Form focuses on abstraction in order to be very clear about what is being inquired about. Peirce called abstraction *prescision* (not to be confused with precision, though the two are not unrelated), and distinguished it from the two other modes of mental separation which he termed "discrimination" (later: distinction), and dissociation. Prescision is the action required to objectively identify Peirce's categories of First, Second, and Third through observation. A fuller explanation of prescision is given in the Glossary.

As a mode of mental analysis, distinction can do no more than identify difference. For those who believe that facts speak for themselves or that observation is value free, nothing more than distinction is considered necessary. Dissociation, for its part, allows one to disregard all that is not of concern in the cacophony. "The proper way of abstraction is by prescinding" (Hausman 1993: 134). Prescision allows for focus on what the medieval scholars called the *quidditas* or "whatness" (*primity*), the *haecceitas* or "thisness" (*secundity*), and the *entitas* or "beingness" (*tertianity*)—the Firstness, Secondness, and Thirdness of concern.

In 1898, in a public lecture, William James was the first to use the term "pragmatism" to denote a newly emerging philosophical position which he acknowledged as being founded by his friend Peirce. Following this lecture, Peirce felt the need to differentiate his position from others—including James—who were using the term to propound a position that, in significant ways, was not aligned with his own. He had this to say:

Pragmatism is a method in philosophy. Philosophy is that branch of positive science (i.e., an investigating theoretical science which inquires what is the fact, in contradistinction to pure mathematics which merely seeks to know what follows from certain hypotheses) which . . . contents itself with so much of experience as pours in upon every man during every hour of his waking life. The study of philosophy consists, therefore, in reflexion and pragmatism is that method of reflexion which is guided by constantly holding in view its purpose and the purpose of the ideas it analyzes, whether these ends be of the nature and uses of action or of thought. (CP5.13 Fn P1 Para 1/6, 1907)

He later adopted the name "pragmaticism" to differentiate his method from the many others—including William James and Ferdinand C. S. Schiller—who had adopted the name pragmatism for their own positions. As pronounced by Peirce: "[T]o serve the precise purpose of expressing the original definition, he begs to announce the birth of the word 'pragmaticism,' which is ugly enough to be safe from kidnappers" (Peirce, 1905a). This could be judged as merely semantics—meaning in language—but his theory of reality went beyond linguistics. Unlike the pragmatism and radical empiricism of William James, which could be said to be a theory of truth, Peirce's pragmaticism is a theory of meaning in reality. For Peirce, "[t]he opinion which is fated to be ultimately agreed to by all who investigate, is what we mean by the truth and the object represented in this opinion is the real" (Peirce, 1878). While he saw truth as something to be found in the long run, his method of inquiry focused on clarity of meaning in reality.

I think it would assist in grasping Peirce's unique vision and the idea of "mattering" which follows from it, to consider, just for a moment, how we process ideas. Jean Piaget [1896–1980], one of the most influential researchers in the area of developmental psychology during the twentieth century, described two processes used by individuals in their attempt to adapt assimilation and accommodation. Piaget was originally trained in the areas of biology and philosophy and considered himself a "genetic epistemologist." He described assimilation and accommodation as complementary processes through which awareness of the outside world is internalized. Although one may predominate at any one moment, they are inseparable and exist in a dialectical relationship. In assimilation, what is perceived in the outside world is incorporated into the internal world without changing the structure of that internal world, but potentially at the cost of "squeezing" the external perceptions to fit. In accommodation, the internal world has to accommodate itself to the evidence with which it is confronted and thus adapt to it. In reality, both are going on at the same time, although most of the time we are assimilating familiar material in the world around us, nevertheless, our minds are also having to adjust to accommodate it.

Because Peirce is unique among thinkers, understanding his method pragmaticism requires accommodation in large measure. Those scholars who attempt to assimilate his work rather than accommodate it can miss the mark and dismiss him out of hand or pronounce him unintelligible or promulgate distorted interpretations of his method.

You may well wonder as you read: What is she getting at? Why is she telling me this? What is she trying to say? Where is her argument? I ask for your forbearance. As a means of disclosure, I have the English language at my disposal. Even with the addition of Peirce's neologisms—many of which have never entered the lexicon and those which have are often misrepresented—I am tied to a linear process. My argument, perforce, presents as a linear process, but one which, contrary to the hegemonic idiom of Western philosophy, is woven of multiple strands. The best analogy of this that I can think of is a plait which, like Peirce's irreducible categories, is a braiding of three tresses of hair, each tress comprising multiple strands of hair.

Poetry also uses language, is linear and depends on semantics, syntax and pragmatics to embody it, but unlike the language of "non-fiction," its meaning is found when it is considered in its wholeness. The forms of poetry are numerous ranging from the highly structured to what is called "free verse." Let me give you an example of *haiku*, a structured form of poetry originating in Japan and if classic conforming to "rules" as follows:

- Japanese *haiku* traditionally consist of seventeen *on*, or sounds, divided into three phrases: five sounds, seven sounds, and five sounds. English poets interpreted *on* as syllable.
- The Japanese word *kiru*, which means "cutting," expresses the notion that *haiku* should always contain two juxtaposed ideas.
- A reference to the season or changing of the seasons, referred to in Japanese as *kigo*, is an essential element of *haiku*.

Let me exemplify this with my own haiku:

Casting on stitches
Algorithms on needles
Winter's solution

My poem is made up of parts and complies with the rules of classic form; however, it is only meaningful when considered as a whole. In this it is like Peirce's architectonic. The poet's purpose, unlike Peirce's, is not intended to articulate a method for appeasing doubt, nor to communicate ideas. Rather, it is to share the reality of experience—to correspond to sentience.

I think it safe to say, however, that for both the mathematician and the poet, intelligibility is vital to the product of their labor—to its telos which is "mattering." I will argue further that value is a necessary condition of intelligibility. In this I hold with Carl Hausman who argues in his paper "Values and the Peircean Categories" (1979):

> Intelligibility would be blind without value, for it would be nothing but a collection of generals. It would have no context by virtue of which generals could be assessed and brought under the control of a condition for their interrelations. (Hausman: 1979, 221)

It is my contention that discerning "mattering" enables objective identification of value in inquiry and action not merely as nominal statements of values or proclaimed ethical codes but as real qualities permeating processes. Through my examples I hope to reveal the apodictic nature of value and thereby extend Peirce's method in which he shows how to make our ideas clear. I argue that value, in being inextricable from process, is both intelligibly and objectively identifiable. My hypothesis of "mattering" is framed as three parts, each part focusing on each Firstness, Secondness, and Thirdness. Ultimately, "mattering" is irreducible to any one or two of the three; it is both an argument and a concept whose reality can be tested in terms of the Pragmatic Maxim.

The universal conditions of interrelations are what Joseph Raz called in his 2001 Tanner lecture, "The Practice of Value" (2003), "enabling or facilitating values." I have identified "beingness" as the telos of "mattering" and three of many values, integrity, respect, and transparency, as conditions of universal interrelations.

It is my hope that my exposition will reveal and validate my three-part hypothesis of "mattering" that

1. value functions as a condition of intelligibility, that telos as the ground of "mattering" is dependent on value for its realization;
2. power—where power is the capacity to cause—is the enabler of force functioning as actual "mattering"; and
3. "mattering" is evolutionary realization of universal telos.

I begin, in chapter 1, with answering: *Why Peirce?* When it comes down to it, it is because Peirce's *pragmaticism* is the only method I have come across for comprehensively and meaningfully exploring questions about "what is going on" and about "how"; about the meaning of "mattering." Chapter 2, Truth *en futuro*, is about reality as an ongoing historical process. Chapter 3, *Belief*, is concerned with the difference between assertion and belief, the relationship between belief and doubt, and finally why understanding rather than truth is the aim of inquiry. Because I am still developing my understanding of power, I infer it. Chapter 4, *Information*, is an exploration of the concept of information, seen by Peirce as the product of matter and form. Because information, *qua* information, is a newly emerging focus for philosophers and scientists, I lean heavily on direct quotation—some may say "slabs"—rather than running the risk of misinterpretation. Chapters 5 and 6 show "mattering" as temporal eventfulness co-dependent with values and infused with power. Because I develop my theory through Peirce's method, chapter 5 begins *"In the Beginning"* with cosmogony, theories or accounts of the origin (L17) and creation (M18) of the universe, followed by chapter 6 where I consider *Cosmology*, theories or postulate accounts of the evolution and structure of the universe (M17) and the branch of philosophy which deals with the universe as a whole (M18). I conclude this chapter by exploring the idea that even the greatest clarity of meaning may continue to leave "mattering" obscured and identify the most recent cosmological discoveries, dark matter, and dark energy as, perhaps, allowing for a fuller disclosure of cosmogony. The core of my theory of "mattering" is covered in chapter 7, *Value and Purpose*, and chapter 8, *Value and Power*, with an exploration of the meaning and co-existence of value as the ground of "mattering," and of how value permeates the process of "mattering." The conclusion is concerned with the consequences of accepting the theory of "mattering" and takes the form of a yarn—which is a tale or a story—of the self-creation of the universe out of freedom and the evolution of its laws through both "consentment" and the discipline of limit of power. It tells what it means in reality to say: It matters; it is "mattering."

Chapter 1

Why Peirce?

I found Peirce quite by chance. I was working in the much-maligned role of "administrator"—senior manager—responsible and accountable for, inter alia, workplace design, planning, and change management at a university center. I was attempting to design robust tools for mapping the operation of quality in scientific research.

Trawling the web, my attention was caught by a paper by Chong Ho Yu, "Abduction? Deduction? Induction? Is There a Logic of Exploratory Data Analysis?" presented at the 1994 Annual Meeting of the American Educational Research Association. The process as described by Yu suggested an opening out of the mind.

> Exploratory data analysis, which aims at suggesting a pattern for further inquiry, contributes to the conceptual or qualitative understanding of a phenomenon. . . . Abduction, the logic suggested by Peirce, can be viewed as a logic of exploratory data analysis. For Peirce abduction is the firstness (possibility, potentiality); deduction, the secondness (existence, actuality); and induction, the thirdness (generality, continuity). Abduction plays the role of generating new ideas or hypotheses; deduction functions as evaluating the hypotheses; and induction is justifying of the hypothesis with empirical data. (Yu 1994: 6)

Yu had presented Peirce as a systematic philosopher who was also a mathematician and scientist and whose model of inquiry suggested that qualitative research was possible. Furthermore, his system appeared to turn epistemological issues into methodological ones—into effort. This, I learned, is what defines pragmatism. I'd been to countless workshops on quality control and "value added" but learned nothing new about quality or value. This was new and exciting.

Growing up in Australia with English immigrant parents and colonial schooling, I was not introduced to American pragmaticism or continental philosophy. Not until college was the idea of process vis-à-vis content high-lighted. Nothing in my schooling told me, and little is said in the literature about the framing of hypotheses, for example, yet there is a great deal written after the fact about inferences.

I read everything by and about Peirce I could lay my hands on. Along the way I learned that Peirce built his philosophy using what he described as the "multiform argumentation of the middle-ages" (CP5.264, 1903). He believed this to be the method of "the successful sciences." A philosophy using this method will

> trust rather to the multitude and variety of its arguments than to the conclusive-ness of any one. Its reasoning should not form a chain which is no stronger than its weakest link, but a cable whose fibers may be ever so slender, provided they are sufficiently numerous and intimately connected. (Smyth 2002: 283)

Let me introduce, to those of you not familiar with him, this American poly-math. Charles Sanders Peirce [1839–1914] was, by profession for most of his working life, a scientist; by inclination, a philosopher in general and a logi-cian in particular; by accident of birth, a male born into a nineteenth-century northern American elitist white family; by nurture and under the influence of his father, Harvard professor Benjamin Peirce, a mathematician. With the exception of a brief stint as a lecturer in logic at Johns Hopkins [1879–1884], he was by profession a scientist, specifically a geodesist with the US Coast Survey and an astronomer at the Harvard Observatory. Until such time as his responsibilities to his employers ceased, it was on his own time that he read and wrote in philosophy. In addition, he conducted an unstinting critique of his writings, making many drafts of his papers to achieve a clarity with which he could be satisfied. He paid the same close attention to his reading and was scrupulous in acknowledging his sources. Despite the fact that he recorded some unflattering first impressions of authors—initially he spoke with con-tempt of Hegel—he ultimately made thorough assessments, affording each (including Hegel) the thoughtful respect often denied him.

He was a voracious reader. In addition to the 8,000 books, many inter-leaved with notes and annotations, which remained at his death, and he had owned many others, including rare titles, which he had sold to support himself. Others he had borrowed from various sources including university collections. He read many in logic, as a way of auditing what texts were being promulgated as right thinking up to and during his lifetime most of which have since gone into obscurity for want of an audience prepared to accept the mediocre. He was fluent in a number of languages including Greek, German, French, and Latin, knew some Egyptian and Arabic, and read texts, where these were available to him, in the original. Apropos Peirce's reading, Joseph Esposito makes an apt comment:

> Peirce studied the history of epistemology as an ethnologist studies a foreign culture, as an outsider trying to understand and make sense of an activity that appears at first blush to be uncomplicated, natural, spontaneous and yet on reflection largely unintelligible. (Esposito, 1998)

This more abstract science, to which Peirce was alluding, metaphysics, he saw as being in a deplorable condition. Having determined metaphysics as a prerequisite to the special sciences, Peirce devoted many years to developing it. Cornelis de Waal in his 2005 paper, "Why metaphysics needs logic and mathematics doesn't," points out that for Peirce

> the issue is not whether we should have a metaphysics—as everyone has a metaphysics whether they want to or not—but whether we want to keep our metaphysics unconscious or bring it out in the open where it can be subjected to the same scrutiny as our scientific work. (de Waal, 2005: 293)

However, he reminds us:

> Never to settle *a priori* what can conceivably be settled by experience. The routine violation of this principle has made metaphysics a discipline that hampers science rather than helps it and has given metaphysicians such a bad rap among scientists. (Ibid.: 294)

Certainly, mathematics is indispensable to special sciences. As de Waal points out:

> The mathematician's main business is . . . the simplification of complicated sets of facts by reducing them to a shape that facilitates their study while still being representative. For this one needs recourse to the three key mental qualities Peirce associated with doing mathematics: imagination, concentration and generalization. (Ibid.: 295)

Yet, while mathematical modeling optimizes necessary reasoning, it does not allow for the *logica docens* required to develop and clarify the *concepts* of metaphysics that, in turn, provides "a roadmap for the special sciences by developing a general system in which all possible facts can be given a place" (ibid.: 294). Yet, over the past three centuries, special scientists have increasingly bypassed philosophy—by which they usually mean metaphysics—seeing it as irrelevant and turning their attention, instead, to mathematics. Paul Thagard (1990) points out that physics split off from philosophy in the seventeenth century, biology in the nineteenth, psychology around the beginning of the twentieth, and linguistics in the mid-twentieth century.

In Peirce's classification:

> Metaphysics may be divided into, i, General Metaphysics, or Ontology; ii, Psychical, or Religious, Metaphysics . . . and iii, Physical Metaphysics. . . . The second and third branches appear at present to look upon one another with supreme contempt. (EP2. 260, 1903)

Philosophy, particularly, metaphysics, is the discipline of inquiry to which *I* am most drawn: the ancients, scholastics, rationalists, empiricists, idealists, commonsense realists, materialists, idealists, pragmatists, process philosophers, phenomenologists, existentialists, and post and neo this-and-that. Although I have learnt much from many of these, I was unconvinced by their metaphysics in cases where it was articulated. I was very drawn to existentialism especially where it emphasized responsibility as related to radical freedom but most of these writers were silent on metaphysics. When I happed upon the thinking of Peirce, I knew I had found a way of comprehensively and meaningfully exploring ideas and qualities and of "mattering."

Around the time the Classic Pragmatists were flourishing, Analytic philosophy, as developed by a number of early twentieth-century British and German luminaries, was colonizing much of the West. What began in opposition to idealism has mutated over the century since its beginnings, infiltrating other positions such as classic pragmatism, and becoming hegemonic. As John Searle notes:

> Analytic philosophy is the dominant mode of philosophizing not only in the United States, but throughout the entire English-speaking world, including Great Britain, Canada, Australia and New Zealand. It is also the dominant mode of philosophizing in Scandinavia and it is also becoming more widespread in Germany, France, Italy, and throughout Latin America. (Searle, 2003: 1)

With the analytic turn, however, many philosophers have made their name. Nevertheless, as philosophy has shrunk, it appears to be seen as less relevant with reactionary, fundamentalist, ideological dogma, both religious and political, too often, being taken as given. The Austrian-British philosopher and linguist Ludwig Wittgenstein [1889–1951] was pronounced by many as arguably the greatest philosopher of the twentieth century. His work marked the breakaway of linguistics noted earlier.

Three-quarters of the way through the twentieth-century Australian philosopher D. M. Armstrong published a pair of slim volumes: *Universals & Scientific Realism,* (1978); *Volume I Nominalism & Realism* rejecting nominalism and *Volume II A Theory of Universals* arguing for scientific realism. Armstrong, although accepting Peirce in large part, ultimately rejected his scientific realism on the grounds of what he called his "Relational Realism." Unfortunately, his misunderstanding of Peirce may well have resulted from his available source of Peirce's work being limited to the *Collected Papers of Charles Sanders Peirce*, Volumes 1–6, edited by Charles Hartshorne and Paul Weiss (1931–1934), and Volumes 7–8, edited by Arthur Burks (1958). The only indication of the editorial policy of the Collected Papers is given in the Introduction to Volume 7:

> The present editor has continued their [Hartshorne and Weiss's] practice of publishing only parts of some of the works, omitting large portions altogether. He has also continued the policy of selecting or compiling a draft for publication whenever there were several drafts available. The present editor has also continued the plan of organization pursued in the previous volumes, breaking up manuscripts, books and series of articles and arranging the resultant materials primarily by subject matter rather than chronologically. (Burks, 1958)

In 1989 at the Charles S. Peirce Sesquicentennial International Congress at Harvard University, a paper "Peirce as Participant in the Bohr-Einstein Discussion" presented by Peder-Voetmann Christansen of the Institute of Mathematics and Physics, Roskilde University Centre in Denmark, concluded that "the future may show that neither Bohr nor Einstein but Peirce had the correct interpretation of a theory he did not even know" (1993: 232).

The American quantum theorist David Bohm, who held Niels Bohr's views on talking about sub-atomic physics, believed that at the root of the problem is the structure of the languages we speak. European languages, he noted, perfectly mirror the classical world of Newtonian physics. Recognizing this, he and his close associate the theoretical physicist David Peat traveled with other leading physicists to meet with

> elders from a number of native American elders of the Blackfoot, Micmac and Ojibwa tribes, all speakers of the Algonquian family of languages. . . . The world view of Algonquian speakers is of flux and change, of objects emerging and folding back into the flux of the world. There is not the same sense of fixed identity. (Peat, 2008)

If one considers classic (i.e., bivalent) logic as a sub-system of the broader category, reasoning, then one can see it as evaluative. Only in the atemporal form that is classic logic might valid argument be deemed true, but so saying it is not a pipeline to truth. The relationship of truth and validity is given by Peter Suber in his course notes "Symbolic Logic."

> Truth and validity are combined in the concept of *soundness*. An argument is sound if (and only if) all its premises are true *and* its reasoning is valid;

all others are unsound. It follows that all sound arguments have true conclusions. . . . Empirical scientists . . . tell us whether statements are true. Logicians
tell us whether reasoning is valid. . . . An argument is valid in a weak sense if it
simply is not invalid. This weak sense of validity turns out to suffice for all the
purposes of rigorous reasoning in science, mathematics, and daily life. (Suber,
1997)

It is its weak sense that Peirce calls abduction. For Peirce, mathematics is the
discipline in which to conduct such an activity. Although he did designate
abduction as logic, in his broad sense of logic, as semeiotic, he considered
it the weakest form. This meets Suber's criteria for soundness but not his
separation between empirical scientists and logicians. Peirce, however, did
not restrict observation to the empiricists but rather saw observation as an
essential activity of all inquiry including the development of hypotheses.

In very simple terms, semeiotic is the science of representations. Representation is a relation of one thing—the *representamen*, or sign—to another—the
object—this relation consisting in the determination of a third—the *interpretant* representamen—to be in the same mode of relation to the second as
the first is to that second, based on some of the leading principles of phenomenology. The presentative aspect of a sign can only be combined with
representative aspects which are equal to or lower than the presentative's
phenomenological type; similarly, the representative aspect of the sign can
only be combined with interpretative aspects which are equal to or lower
than the representative's phenomenological type. This reduces twenty-seven
possible classes of signs to ten.

His method, in which empiricism and rationalism are codependent, aims
for meaning rather than the elusive certainty of truth. Peirce came to realize, nevertheless, that achieving even fallible validity required more than a
hypothesis, empiricism, and rationalism to settle genuine doubt. Metaphysics
is essential to the process.

If we think of metaphysics as the network of theories, we build of the
"how" and the "what" of the universe—of how the developing parts fit to and
with the evolving whole—we will come to realize that it is essential for reality checking. Ethics, of which logic is a kind, is the actualizing. Aesthetics,
which informs ethics, is concerned with what may be fitting, and phenomenology is the means of focusing and orienting inquiry. This makes sense to
me: not "what's that," so much as "what's going on." Why Peirce indeed.

It took Peirce his lifetime to develop his method of achieving intelligibility,
but as order is brought to the disarray of his work, the sheer elegance of his
detailed architectonic is being realized. Peirce, like Kant, adopted the term
"architectonic" to describe his system. He found it an appropriate metaphor,
as he wrote in "Proem: the architectonic character of philosophy":

A great building . . . is meant for the whole people and is erected by the exertions of an army representative of the whole people. It is the message with which an age is charged and which it delivers to posterity. Consequently, thought characteristic of an individual—the piquant, the nice, the clever—is too little to play any but the most subordinate role in architecture. (CP1.176, 1893)

The Sydney Opera House comes to mind for me in finding this description so apt. Jørn Utzon, the original architect, is quoted as having said:

I have made a sculpture . . . you will never be finished with—when you pass around it or see it against the sky . . . something new goes on all the time . . . together with the sun, the light and the clouds, it makes a living thing. (Utzon)

Those who have read this undated quotation, in many places it is reproduced, would most likely think that it refers to the actual building. It is, in fact, from the text accompanying the original design sketch submitted in 1957 and relates to what was then the germ of an idea. Among other things, it illustrates exceedingly well aspects of Peirce's thinking underlying his classification of the sciences. Figure 1.1 (below) is a reproduction of this sketch.

Before embarking on details of this classification, it is necessary to throw some light on his thinking in relation to "ideas" and, via "definition," "science."

First, "ideas," he wrote:

If you ask what mode of being is supposed to belong to an idea that is in no mind, the reply will come that undoubtedly the idea must be embodied (or ensouled—it is all one) in order to attain complete being and that if, at any moment, it should happen that an idea—say that of physical decency—was quite unconceived by any living being, then its mode of being (supposing that

Figure 1.1   Jørn Utzon's Sketch of His 1956 Winning Submission to the Sydney Opera House Design Competition. *Source:* The Red Book, 1958, p. 1. (© Mitchell Library, State Library of New South Wales and Courtesy Jørn Utzon.)

it was not altogether dead) would consist precisely in this, namely, that it was about to receive embodiment (or ensoulment) and to work in the world.

What I do insist upon is not now the infinite vitality of those particular ideas, but that every idea has in some measure, in the same sense that those are supposed to have it in unlimited measure, the power to work out physical and psychical results. They have life, generative life. (EP2. 123, 1902)

How ideas are embodied, considered, and applied is revealed in Peirce's perennial classification of the sciences which is the blueprint of his architectonic. Beverly Kent's *Charles S. Peirce: Logic and the Classification of the Sciences* (1987) is the most comprehensive and thorough secondary source of this and it is her research that brings me here. Kent surveys Peirce's examination of previous attempts throughout history to classify the sciences and his own attempts from his 1866 Lowell lecture through to the period 1903–1911 when it stabilized. The most influential earlier attempt for him was that of Auguste Comte [1798–1857] though it is wrong to think, as some did in the early part of the twentieth century, that he was thus a positivist. To understand what Peirce meant by qualifying "science" as "positive" in his classification, one needs to realize his definition of "definition."

The definition and the utility of a definition require it to specify everything essential and to omit all that is inessential, to its *definitum*: though it may be pardoned for calling special attention to an omission in order to show that it was not inconsiderate. (EP2. 454, 1911)

In an earlier paper, he brought to our attention that "[a] definition does not assert that anything exists" (EP 2. 302, 1904). Peirce found both method and system essential to the definition of science, yet neither of these convey its "livingness," nor whether it is dependent. Although Coleridge's *Encyclopædia Metropolitana* (1852) defined science as systematized knowledge, it did so without suggesting that the organization be in accordance with principles (see CP7.54, c.1902). Furthermore, it included what Peirce referred to as "fossilized remains" of science (MS.614, 1908).

This classification, which aims to base itself on the principal affinities of the objects classified, is concerned not with all possible sciences, nor with so many branches of knowledge, but with sciences in their present condition, as so many businesses of groups of living men. It borrows its idea from Comte's classification; namely, the idea that one science depends upon another for fundamental principles, but does not furnish such principles to that other. (EP2. 258, 1903)

Science is inquiry which is systematically directed toward developing ideas and discovering truth. Positive science is experiential and is principle dependent within a natural classification. Peirce saw that any two sciences could be related in three different ways: by the relationship of material content, dynamical action, and rational government. Having no illusion that any one scheme could capture all of the relations of dependence, he opted for the third ordering as being the most stable. Kent points out:

> Peirce maintained that, with the single exception of mathematics, every science employs without question a principle discovered by some other science. The latter science may call upon the narrower one for data, problems, suggestions and fields of application. (Kent, 1987: 124)

Understood in terms of the activities of those who are directed to developing ideas and truth, science falls into the category of mind; each of the three branches, heuretic sciences, sciences of review, and practical sciences, is distinguished by its different motive. Without going into detail here or into all Peirce's particular meaning of terms, what follows is a summary of his classification of the sciences.

Heuretic, meaning (designating or pertaining to) that which treats discovery or invention, is from the Greek *heuretikos*, inventive, which is from *heuriskein*, find *heuretikos*, inventive. Heuretic sciences are motivated by discovery for the sake of discovery and are formal, theoretic, and originate in observation. The differentiating idea is concerned with the relation of the sciences to phenomena either in terms of the kind of phenomena observed or in terms of the kind of assertions made as a result of reasoning on those observations, that is, the sort of orientation each heuretic science takes toward developing ideas and truth.

Mathematics is first in the classification and is so because it is the only science that is not principle dependent. The differentiating idea of mathematics concerns the general nature of the hypotheses it creates, which are distinguished by the multitude of elements hypothesized and then in terms of the relations between those elements.

Philosophy, the second heuretic science, inquires into positive universal truth using principles discovered in mathematics. It examines the phenomenon of ordinary experience in terms of its mode of being. Utilizing the familiar experience acknowledged by everyone, observations are of the phenomena common to all. Philosophy's sub-classes, or orders, are, in the first place phenomenology, in the second, the normative sciences, being aesthetics, ethics, and logic, and in the third place, metaphysics.

Phenomenology focuses on what Peirce called the phaneron as it immediately presents itself. It provides the observational groundwork for the rest of philosophy, endeavoring to determine the universal indecomposable elements in whatever appears before the mind.

Normative sciences study the phenomenon insofar as we can act upon it and it on us and endeavor to determine the conditions required for an object to be fine irrespective of whether any specific objects possess that fineness. The three sub-classes are esthetics, ethics, and logic. Esthetics (to use Peirce's spelling) inquires into the deliberate formation of habits of feeling that are consistent with the aesthetic ideal. Ethics inquires into the theory of the formation of habits of action that are consistent with the deliberately adopted aim. Logic studies the deliberate formation of habits of thought that are consistent with the logical end.

Metaphysics, using principles of logic, inquires into what is real (and not figment) as far as can be ascertained from ordinary experience. It is differentiated by the various relations in the different kinds of phenomena discovered to be real.

The third heuristic science comprises the special sciences—those most commonly understood as science. Entailing observations of previously unknown phenomena, this class is divided by Peirce into physics and psychics—meaning the physical sciences and the human sciences.

The sciences of review are motivated by discovery for the sake of applying knowledge. This is the class under which this review of pragmaticism began.

All the sciences to this point in the hierarchy may be considered theoretical. The final class in Peirce's classification is the practical sciences where discovery is for the sake of doing. This class, significantly, is differentiated from those above by its concern with ulterior purpose.

Strong parallels with Peirce's evolutionary view are expressed in the article "The flexi-laws of physics" by the physicist Paul Davies (*New Scientist* June 30, 2007b). Davies begins: "Science works because the universe is ordered in an intelligible way." He concludes:

> In the orthodox view, the laws of physics are floating in an explanatory void. Ironically, the essence of the scientific method is rationality and logic: we suppose that things are the way they are for a reason. Yet when it comes to the laws of physics themselves, well, we are asked to accept that they exist "reasonlessly." If that were correct, then the entire edifice of science would ultimately be founded on absurdity. By bringing the laws of physics within the compass of science and fusing nature and its laws into a mutually self-consistent explanation, we have some hope of understanding why the laws are what they are. In addition, we can begin to glimpse how we, the observers of this remarkable universe, fit into the great cosmic scheme. (Davies, 2007b)

There is a hidden assumption here that applies to all the aforementioned and all the ones discussed, including mathematics: *power*. Power is ubiquitous—to say or even think otherwise is an oxymoron.

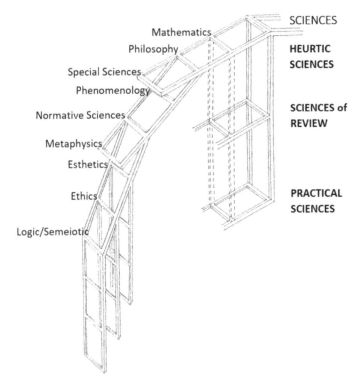

Mathematics
Philosophy
Special Sciences
Phenomenology
Normative Sciences
Metaphysics
Esthetics
Ethics
Logic/Semeiotic

SCIENCES

HEURTIC SCIENCES

SCIENCES of REVIEW

PRACTICAL SCIENCES

Figure 1.2 Ladders Attached to the Normative Sciences Rung, Which in Turn Is Attached to the Philosophy Rung on the Heuretic Sciences Ladder Depicting Peirce's Description of a Stereoscopic Image of His Perennial Classification. *Source*: Adaptation of Beverley Kent (1987: 136).

Given that philosophy is only one aspect of Peirce's architectonic, I have adapted a diagram constructed by Kent to depict Peirce's description of a stereoscopic image of his perennial classification (see Figure 1.2 above). This depiction shows it as a model for learning and all that entails. She wrote of it as follows:

> He [Peirce] left a verbal description only but it is known that he thought in three-dimensional diagrams and that he participated in the early development of lattices. (Kent, 1987: 135–6)

Why Peirce? At the outset, I put forward the question: What does it mean to say something, anything, everything matters, and what is *it* that matters— what *is* "mattering"? I chose Peirce because he thought temporally and, in 3D, utilizing his method of inquiry and seeing it as codependent on value, I can explicate "mattering." But first…

Chapter 2

Truth *en future*

Peirce was not blind to the inadequacy of either empiricism or rationalism on their own for discovering the truth. Exploration needs to begin with creativity, imagination, theory, and hypothesis. Here it is worth visiting Peirce's theory of "true" as a quality of the three kinds of representations: verisimilitude, veracity and perfect veracity, or verity. Quoting his argument from his 1861 "Treatise on metaphysics":

> The objection to verisimilitude's being the truth of conception is its limitation as to completeness; the objection to veracity's is its limitation in beginning. Neither is open to the objection against the other. But veracity was called that kind of truth which was not verisimilitude. Conceive, however, veracity to be perfect—to be founded not upon convention but upon the very nature of things and what have we?

1. The nature of a thing is that which it derives from its origin. Derivation not in time is the relation of accident to substance. Hence, an invariable connection in the nature of things is unity of substance.
2. The qualities of things are founded in the nature of things; hence, unity of substance implies perfect correspondence of qualities.
3. Hence perfect veracity is of a distinct character from cognizable veracity and it approaches quite as nearly perfection of verisimilitude. I will call it "verity" and the representation a "type."
4. Since conceptions perfectly correspond with qualities and since they have a connection therewith in the nature of things, they are "types" of things. (W1. 80, 1861)

Verity, however, is not *directly* verifiable—there is nothing more than all with which to validate it. Peirce had it that inquiry is initiated when something surprising occurs to bring some belief into doubt. Given there is no

21

certainty *en futuro*, then, for the sake of clarity, when we engage in inquiry, we need to be upfront about our assumptions before launching into inquiry. As Peter Skagestad tells us:

> From the vantage point of the observer the notion of truth as correspondence is adequate (although it will no longer do when applied to his own beliefs). The scientist needs an equally absolute notion of truth as a regulative idea. (Skagestad, 1981: 83)

What *is* achievable, if we are prepared to accept fallibilism is cognizable veracity (CP1.14, 1897).

While he accepted that all reality is not known, such acceptance is not tantamount to conceding to incognizability. Peirce did not countenance the inexplicable, seeing such an idea, exemplified, for instance, in Kant's "thing-in-itself," as incoherent. In her book *Peirce's Pragmatic Theory of Inquiry: Fallibilism and Indeterminacy* (2007), Elizabeth Cooke points out that "the commitment to fallibilism is a commitment to the belief that there are knowable truths about which inquirers are fallible" (p. 23). She speaks of Peirce's model of knowledge as "adaptionalist" noting that on his account, "Knowledge is not 'true belief' but is the integration of rational habits with ontological habits" (p. 28.)

Peirce's method, used as a tool for obtaining reflective clarity of intellectual concepts and propositions, can produce the finely honed validation necessary for developing a cognizable metaphysics. When Peirce first proposed his Pragmatic Maxim, he explained how it was to be applied to the doctrine of reality. He gave the etymology and a definition of "real" in his paper "A Neglected Argument to the Reality of God" (NA).

> "Real" is a word invented in the thirteenth century to signify having Properties, i.e. characters sufficing to identify their subject and possessing these whether they be anywise attributed to it by any single man or group of men, or not. . . . The "Actual" is that which is met with in the past, present, or future. (Peirce, 1908)

Peirce did not explicate "mattering" directly, maybe because, as he saw it, the purpose of a natural classification is to render phenomena comprehensible. He believed that "form is something that the mind can assimilate and comprehend, while Matter is always foreign to it, a recognizable but incomprehensible something" (MS.499(s), 1907). This may be the case if matter is viewed as a *fait accompli* but not when considered as process—the process of "mattering." Process, in turn, assumes power. In his book *Representative Practices*, Kory Sorrell points out:

Peirce did not have an ontological category of process. There is quality, brute actuality and continuity or lawful behavior, but not process *as such*. Yet, for Peirce, this set of categories does not signify the irrelevance of process, but rather its absolute certainty. (Sorrell, 2004: 34–35)

In his early paper "On a New List of Categories" (1867), Peirce "refers to his categories as conceptions necessary to unify experience through joining subjects and predicates" (Hausman, 1979: 204); by 1894, he had developed the categories on the basis of his logic of relations.

> Following this logic, the categories are defined as three kinds of relation, each relation being defined by the number of its relata. From this perspective, the categories are most appropriately called "the monad," "the dyad," and "the triad." (Hausman, 1979: 204)

These are his categories from the standpoint of the logic of relations. Peirce, however, identified and described his categories from two perspectives, one formal or logical and one material or phenomenological. From the perspective of his phenomenology, he called his universal categories Firstness, Secondness, and Thirdness. "As given in phenomena," Hausman reminds us, "they are empirically discoverable; yet they also condition what is discovered" (ibid.).

As commonly understood to say that phenomena are empirically discoverable is to say they can be perceived. The experience of perception—"that knowledge which is directly forced upon you and which there is no criticizing, because it is directly forced upon you" (CP2.141, 1902)—is what Peirce came to call *haecceity* (Latin for "thisness"). This term *haecceity* is one he adopted from Duns Scotus [1266–1308] and adapted to signify the existence of quantifiers of "thisness" or "hereness and nowness" and thereby unambiguously distinguish the particular.

> It is existent, in that its being does not consist in any qualities, but in its effects—in its actually acting and being acted on, so long as this action and suffering endures. Those who experience its effects perceive and know it in that action; and just that constitutes its very being. It is not in perceiving its qualities that they know it, but in hefting its insistency then and there, which Duns calls its *haecceitas*. (CP6.318, 1908)

This appears, at first glance, to fly in the face of Peirce's radical realism, in which, as he puts it "the absolute individual can not only not be realized in sense or thought, but cannot exist, properly speaking" (CP3.93n, 1870). Second glance shows this is not the case. An *absolute* individual would be a monad or First which is no more than a possibility; it can only come into existence by being embodied or determined through Secondness which is a dyad, that is, not an *absolute* individual. As an experience of a dyad, Jeffrey DiLeo (1991) stresses that "*haecceity* is perceived and not inferred" (p. 97), that is, it is "brutal fact" and not Scotus's contraction, "the process whereby the common nature was converted into an individual possessing numerical unit" (p. 98). DiLeo points out that "Peirce rejected contraction as a process of individuation" but that in doing so "he is merely denying 'reductionism' of the categories" and not particularity. Expanding on Peirce's *haecceity* as the experience of "brutal fact," DiLeo explains:

> For Peirce, because the "facts" are characterized by particularity, they are completely determinate in regard to whether they are in the possession of qualities. Moreover, the principles of contradiction and excluded middle are applicable only to what is individual, viz., "the individual is determinate in regard to every possibility, or quality, either as possessing it or as not possessing it" (CP1.434, 1896). Thus, given that the principles of contradiction and excluded middle hold for *haecceity*, the qualitative experience of *haecceity* is impossible. (DiLeo, 1991: 97–98)

It could be argued that Peirce, in developing his phenomenology, introduced *haecceity* because as DiLeo (p. 98) further notes: "Without individuals and *haecceity* there would be neither Secondness nor perception and without perception Peirce's epistemology would be untenable." Phenomena, however, cannot be reducible to the category of Secondness. *Haecceity* is not reality but rather is abstracted from reality which in the case of Peirce's phenomenology is the phaneron.

The purpose of phenomenology as Peirce defined it is to describe *what* is before the mind. The "whatness" is, utilizing another Scholastic term, the *quidditas* or primanity; it is the Firstness of Secondness. However, identifying the *quidditas* entails, in the first instance, mental separation of the particularity of *haecceity* or "thisness" from the unity of "thusness" of the phaneron. Peirce first discussed the grades of separability of one idea from another in "On a new list of categories" (1867) then raised them again in detailing the doctrine of categories which condition what is discoverable through his phenomenology. In his later writings, he called the three grades of separability "dissociation," "prescission," and "distinction" (rather than "discrimination" as in his 1867 paper) when describing them:

The change by Peirce of early use of "discrimination" to "distinction" may be explained by the shift in the meaning of "discrimination" in his lifetime. Originally its meaning was the action or an act of discriminating or distinguishing; the fact or condition of being discriminated or distinguished; and a distinction made (M17). Later in his life it came to mean *spec*: the practice or an instance of discriminating against people on grounds of race, color, sex, social status, age, and so on and an unjust or prejudicial distinction (*L19*). In this pejorative sense, men and women, for example, are perceived as the *entitas*—beingness—rather than as the haecceitas—thisness—of the *entitas*: people.

As a child, I originally learnt the word "discriminating" from my English grandmother as a word to describe "taste": to have "discriminating taste" was to have what was judged to be "good" taste. Born and raised in the second half of the nineteenth century, she had learnt the *M17* meaning of the word. She was a dressmaker who learnt her trade before the invention of man-made fabric. Through experience and ongoing learning, she came to prefer natural fiber and had taught me how to tell the difference through touch—what she called "feeling the fabric." She also taught me what she had learnt of the differences and why she preferred one over another. What she had taught me was some of the short- and long-term consequences of using one over another and in what circumstances. I did not learn about the *L19* use of "discrimination" until I was in my teens and only over time, the breadth and depth of its reality.

Reference to the design and construction of the Sydney Opera House—in this case the tiles of the "sails" as shown next—can serve to further exemplify the grades of separability of one idea from another.

These sketches of the individual tile parts for assembly of the covering of the "sails" and this drawing of an assembly of tiles were found at the Mitchell Library, Sydney, home of sketches, drawings, photographs, and models of the design of Sydney Opera House.

Apart from the obvious that both could be dissociated from the reproductions shown here, these images cannot be supposed without a medium on which to present them, be it the paper of the originals or any of a variety of media to display the reproductions. Both are of the same parts of the Opera House shells, yet they are distinct because not only do they occupy separate places but figure 2.1 is an assembly of the parts shown in figure 2.2. Furthermore, they are chronologically distinct, the second being an example of an assembly of the first. Anyone viewing the originals would see that both were drawn with pencils as opposed to ink, prescision can reveal that two different pencils were used. Most pencil cores are made of graphite (one of the allotropes of carbon) mixed with a clay (a naturally occurring aluminum silicate) binder, leaving gray or black marks. Graphite is durable: although it can be removed with an eraser, it is resistant to moisture, most chemicals, ultraviolet radiation, and natural aging. Pencils are graded according to the mix of graphite and clay from "B" (for blackness) to "H" (for hardness). Figure 2.2 is blacker than figure 2.1. Yuzo Mikami, who served on the architect team for a while, confirms this. In his book *Utzon's Sphere: Sydney Opera House—How It Was Designed and Built* (2001) he talks of "the battle [*sic*] of B6 and H2 pencils." Utzon, he tells us, used the B6, it is perfect for locating form and rhythm in thick, soft strokes as shown in figure 1.1. The H2 pencils of the technical teams were used to convert the ideas into blueprints. Figure 2.2

Chapter 2

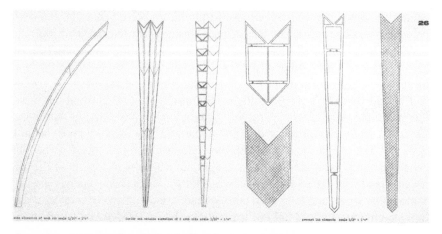

Figure 2.1 Precast Spheroidal Lid-Elements. *Source*: The Yellow Book, 1962, p. 26. © Mitchell Library, State Library of New South Wales and Courtesy Jørn Utzon.

combines these two, with the B6 pencil overlayered to emphasize the fan-like assembly of a portion of the shell.

DiLeo notes: "For Peirce, not only is *haecceity* only known through experience, its very being is determined through experience" (1991: 93). Considered as a process, Peirce's categories reveal the pervasive features of experience by phenomena. When we engage in his phenomenology, we are really experiencing experiencing. Sorrell notes importantly that experience denotes acting *and* suffering.

> But experience also denotes history; it is the storied past that informs the present, provides a context for its explanation and directs future development along some lines and not others. . . . When both senses of the term are operative, experience is seen to be of things that are in the process of development or change, sometimes radically so, but continuous with their respective pasts. (Sorrell, 2004: 39)

Without undermining Peirce's arguments of differences between epistemology and ontology, and causality and causation, as nascent, phenomena are not so much things as events. Put another way, phenomena are eventful. Paper is a good example of this. Paper is produced by pressing together moist fibers, typically cellulose pulp and drying them into flexible sheets. Much of the early paper made from wood pulp contained significant amounts of alum, a variety of aluminum sulfate salts that are significantly acidic. Alum was added to the paper to assist in sizing the paper, that is, making it somewhat water resistant so that inks do not "run" or spread uncontrollably. The early papermakers did not realize that the alum they added liberally to cure almost every problem encountered in making their product would eventually

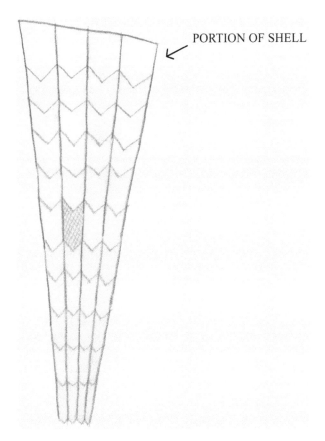

PORTION OF SHELL

Figure 2.2   Detailed View of Tile Lid Panels/Scale: Approx. 1/8″=1′0″. *Source*: © Mitchell Library, State Library of New South Wales and Courtesy Jørn Utzon.

be detrimental. The cellulose fibers which make up paper are hydrolyzed by acid and the presence of alum eventually degrades the fibers until the paper disintegrates in a process which has come to be known as "slow fire." While the use of non-acidic additives in the manufacturing process makes for greater stability of paper, all paper is at risk of acid decay, because cellulose itself produces formic, acetic, lactic, and oxalic acids. Paper is an event.

Furthermore, we as observers are as much phenomena as that which we observe; observers and observed are experienced and experiencing. Crossing roads of moving traffic exemplifies this. Peirce pointed out that it is because we find ourselves in such complicated situations that it is impossible to determine with exactitude what the consequences could be that one calls for the help of the mathematician (N2: 9). Some years ago I realized that crossing city roads, as I had been doing for as long as I could remember and living unscathed to tell the story meant that, contrary to what I had until then believed, I can do mathematics. Once we understand that mathematics is less about manipulating notations and more about considering the relationships embodied in vital phenomena, we realize that we are all mathematicians.

To do the preliminary mathematics whereby I frame the hypothesis with the view to the drawing of necessary conclusions and study what is true of the hypothetical states of things, I need to be able to imagine, concentrate, and generalize. Campos elucidates this.

> In his writings on mathematics, Peirce often emphasizes the abilities of imagination, concentration and generalization that are necessary for mathematical reasoning. . . . In my estimation, these powers correspond, respectively, to the necessary abilities to (i) create a mathematical "icon"—a presentation of a hypothetical state of things that is of interest for its own intrinsic formal character to the inquirer *qua* mathematician; (ii) discriminate between mathematically essential and superfluous relations in the determination of the icon and focus the attention on the essential ones; and (iii) generalize on the basis of the characters and relations embodied in the icon. (Campos, 2009: 137)

Certainly, my general hypothesis, that I can cross busy roads without mishap, has proved true to date but it can never be guaranteed—every occasion is an exercise in theorematic reasoning following from the *haecceity* of the situation. Each time I stop at a curb intending to cross, I again need to study what is true of hypothetical states of things by considering the variables and the relationships embodied therein.

The variables that would impact on my hypothesis include, but are not limited to, the number of lanes in one or two directions of moving traffic and whether traffic is intermittent, continual, and continuous; whether overtaking is allowed; whether there are any roundabouts or whether there are cross streets or T junctions and whether turning into or out of these, left or right, is permitted; whether stopping, standing, and parking is permitted; whether there are parked vehicles and whether there is any indication that any of these may enter traffic; whether there are parking spaces; what kind of and in what condition is the road surface; what time of day is it and therefore what is the position of the light source; what are the weather conditions; and, just as important, because I am implicated in the hypothesis, what is my physical and mental condition at the time.

To perceive the actual state of things, I need to be able to dissociate, prescind, and distinguish. Because this is a complex of events, utilizing Ockham's razor, I dissociate many seconds from thirds, mostly unconsciously and even before I stop at the curb. In terms of weather conditions, for example, I dissociate fog. If, in fact, fog is present, I would not even consider crossing a road because, apart from anything else, essential perception would be compromised. Because momentum is pervasive and because any sudden increase is necessarily precluded from my hypothesis, prescision is only of seconds from thirds. An exception might be if the road had suffered a storm that produced potholes deep enough to create the possibility of changes in the momentum of vehicles driving over them and of water collected in them. Notwithstanding these kinds of events, in terms of traffic, I prescind distances and speed, velocity and acceleration, distinguish left from right, nearer from further and faster from slower, and so on.

I then need to ascertain the best—the aesthetic—juxtaposition of all the relationships of all the variables. To identify the space-time to cross I need to be able to generalize, anticipate, and predict. What I am doing is identifying nascent patterns of the events before me, looking for the best moment to step in becoming another event within it without disturbing the prevailing momentum. If, as I wait to cross, any element of chaos, such as even the smallest traffic accident, breaks the pattern, my hypothesis cannot be realized in that situation and I temporarily relinquish my purpose. If, however, I am confident that the necessary conditions of my hypothesis are prevailing, I cross.

All this I do without the aid of any instruments or traffic controls and without using notations. I cannot tell you the speed of any vehicle or how long it takes to cross a road or any of the distances involved. Nevertheless, observation has taught me that in traversing the vector AD (see figure 2.3) across a one-way, single-lane road as depicted here, that the vectors AB, EB, FB, and GB are moving at different speeds in equivalent time. Only the relationships between the time vector AC and HB are safe.

Certainly, to do this, I had to have achieved the level of cognitive development which Piaget called concrete operation enabling me to mentally manipulate information that is present. While there was much I learnt through

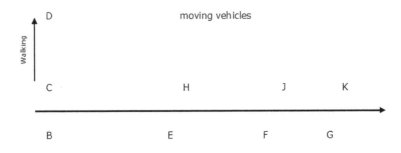

Figure 2.3   Example of Juxtaposition of Events in Space-Time.

experience during that period of my developing *logica utens*, conservation and a decline in egocentrism were perhaps the most important. Through the ongoing processes of assimilation and accommodation, the concepts of invariance, causality, and distance necessary for observing the relationships of events in space and time develop. Observing the relationships of events and generalizing from these, trusting that drivers will more or less obey the man-made rules of the road and with faith that natural laws will hold, I cross if and when I can.

Haecceity enables identification of existence, but being Secondness, "this-ness" is always in the act of becoming "thatness," "thereness," and "then-ness" and therefore can be generalized, anticipated, and predicted.

Only what we call "now" is still. It is analogous to a portal through which possibility streams from nothing into being and becoming. Everything that has been, or is becoming, has passed or is passing through "now." All that has been or is becoming, including space, time, light, matter, ideas, mind, natural laws, moves, and so on. Actuality is eventful. That which has passed through the brute force of "now" has become fact—whether it is discovered or not—and therefore is bivalent but only as fact. "Now" is the difference between the real and reality. Peirce spoke of the real as "that which is not whatever we happen to think it, but is unaffected by what we may think of it" (CP8: 12, 1871). This is only strictly so in the past tense. As a futurist he wrote, "The opinion which is fated to be ultimately agreed to by all who investigate is what we mean by the truth and the object represented in this opinion is the real" (Peirce, 1878). Lest his use of "fated" was taken as meaning something like predetermined, he wrote as a footnote:

> Fate means merely that which is sure to come true and can nohow be avoided. It is a superstition to suppose that a certain sort of events is ever fated and it is another to suppose that the word fate can never be freed from its superstitious taint. We are all fated to die. (Peirce, Fn 2, 1893)

The real as fact is past; reality as truth streams ahead—it is *en futuro*. Identifying facts might be achieved by the individual historian, geologist, and so on, but it takes the community to predict the truth. John E. Smith in his paper "Community and Reality" (1980) speaks of Peirce's notion of community as "the willingness of each individual member to sacrifice what is personal and private to him alone in order to follow the dictates of an interpersonal method that involves free exchange of views and results" (p. 50). In this paper Smith admitted that "in an earlier paper on Peirce [1952] I quite mistakenly described him as *not* being a systematic philosopher"; that in this "I was deceived"; and that "I did confound the form in which his writings have come to us with the logical structure of his thought" (p. 38). Finding error in others is ubiquitous; admission to being wrong is rare. Let me follow with Smith's opening remark:

> Charles Peirce was at once a genuine and a disturbing philosopher. He was genuine because he dealt directly with the difficult problems of philosophy: the nature of truth, the theory of reality, the problem of mind and God. In his approach there was none of the modern or advanced tendency of thought that anticipates speculative problems merely in order to avoid them or dissolve them into the misadventures of human speech. Peirce is disturbing because he forces us to confront experience afresh and if need be to revise our categories so that this will accord with the real world. If you study Peirce you have to be prepared for surprises; you have to be tough-minded enough to consider the possibility that things may in fact prove to be very different from the way you have long since decided that they *must* be. (Smith, 1980: 38)

The findings of inquiry using Peirce's method are provisional—they are intelligible in an ongoing process directed at truth, that is, cognizable veracity. Perfect veracity, that is, truth, may only be arrived at in the long run. Having said this, I should say something about what is meant by speaking of "truth in the long run." Robert Almeder (1985) identified a minimum of thirteen theories of truth in Peirce's oeuvre dating from 1873 to 1905, just three of which he deemed intelligible. Without making reference to any other of the many secondary writers, my interpretation is that Peirce was equivocal on the subject, and differed depending on whether he was focused on inquiry or an object. From the former perspective, his Baldwin Dictionary definition is:

> Truth is that concordance of an abstract statement with the ideal limit towards which endless investigation would tend to bring scientific belief, which concordance the abstract statement may possess by virtue of the confession of its inaccuracy and one-sidedness and this confession is an essential ingredient of truth. (Baldwin, 1902)

His most pithy statement from the latter perspective that I found is that "truth is the correspondence of a representation with its object" (CP 5.553, 1902). The poem π by the 1996 Polish recipient of the Noble Prize for Literature, Wisława Szymborska, expresses this notion together with the meaning of truth in the long run:

π

π deserves our full admiration

three point four one.

All its following digits are also non-recurring,

five nine two because it never ends.
It cannot be grasped *six five three five* at a glance,
eight nine in a calculus
seven nine in imagination,
or even *three two three eight* in a conceit, that is, a comparison
four six with anything else
two six four three in the world.
The longest snake on earth breaks off after several metres.
Likewise, though at greater length, do fabled snakes.
The series comprising π
doesn't stop at the edge of the sheet,
it can stretch across the table, through the air,
through the wall, leaf, bird's nest, clouds, straight to heaven,
through all the heavens' chasms and distensions.
How short, how mouse-like is the comet's tail!
How frail a star's ray, that it bends in any bit of space!
Meanwhile, *two three fifteen three hundred nineteen*
my telephone number the size of your shirt
the year nineteen hundred and seventy three sixth floor
the number of inhabitants sixty-five pennies
the waist measurement two fingers a charade a code,
in which *singing still dost soar, and soaring every singest*
and *please be calm*
and also *heaven and earth shall pass away*,
but not π , no, certainly not,
she's still on with her passable *five*
above average *eight*
the not-final *seven*
urging, yes, urging a sluggish eternity
to persevere.
(Wisława Szymborska [Trans. Adam Czerniawski])

To my knowledge, there is no disagreement in the scientific community—
with anyone or any community of "anyones"—that the truth of π is in its
representation but that though it is contained in the symbol, it is *en futuro*.
Notwithstanding the degree of confidence we enjoy with regard to π, we
need to hold in mind that believing truth to be contained in an abstract
statement means it can only be grasped as a tendency *en futuro*. Saying this
is a "confession of its inaccuracy and one-sidedness and this confession is
an essential ingredient of truth" (CP5.565, 1901). No matter how stable we
deem any tendency, we cannot ascribe certainty to it by reference to it a
priori—looking back is useful insofar as it can render history but so saying,
it does not guarantee certainty ahead of now. We must confess to the fallibility of our conclusions for the very reason that meaning always includes,

besides the Firstness of potentiality and the Secondness of actuality, the Thirdness, of possibility.

My hypothesis of "mattering" certainly requires rigorous exploration and demonstration to reveal the possibility that it is "true" but more importantly, that its *p*-value is significant. Experience—including accumulated experience and knowledge of actuality—only travels the continuum as far as now. I can speak of *knowing* "mattered" and "matters" but not with any certainty concerning the truth of "mattering" which is contained in its Thirdness and yet is irreducible from its Firstness and Secondness. Although truth cannot be claimed as *known* without the rider of fallibility, any such claim can be made intelligible, that is, reason can be shown for such a claim—it can be understood.

Peter Forrest speaks of this as aesthetic understanding.

> The aesthetic interpretation of Science states that there are two desiderata for a scientific theory. The first is empirical adequacy. . . . The other desideratum is that the theory, if correct, reveals the aesthetic value of those aspects of the world with which it is concerned. The term currently used to express this aesthetic value is "elegance." Another value term which might be appropriate in place of "elegance" is "fittingness." (Forrest, 1991: 527)

Without knowing Peirce's work, the physicist Paul Davies, like Peirce, is of the opinion that science is faith based. As Davies writes in the *New York Times* Op-Ed article *Taking Science on Faith* (November 24, 2007), "science has its own faith-based belief system."

> All science proceeds on the assumption that nature is ordered in a rational and intelligible way. You couldn't be a scientist if you thought the universe was a meaningless jumble of odds and ends haphazardly juxtaposed. When physicists probe to a deeper level of subatomic structure, or astronomers extend the reach of their instruments, they expect to encounter additional elegant mathematical order. And so far this faith has been justified. (Davies, November 24, 2007)

Faith follows from a confession of fallibility as a rider to a claim of truth. Saying that "this," "that," "something," "anything," "nothing," "it" matters is saying at least that "it" is substantive, but more usually with the emphasis that "it" is significant or has significance. Being made intelligible through process means unfolding patterns of truth contained in concepts that can be understood and given *p* values by inquirers.

Terry Eagleton has suggested: "Faith is for the most part performative rather than propositional" (2009: 111). Earlier he speaks of "an article of faith" the definition of which given in the Cambridge Dictionaries Online is "something you believe in very strongly" and in the Merriam-Webster as "basic belief." He has to say:

> There are those for whom the spectacular successes of science have rendered religion redundant; and there are others for whom those successes spring from a fundamental fact—that our minds seem somehow attuned to the fundamental stuff of the world—which is itself cause for metaphysical reflection.
>
> Why is it that mathematics of all things, seems to encode the intelligibility of the physical universe, and is it reasonable for science to take this, along with the uniformity of physical laws, simply as an article of faith? Is it equally reasonable for science to place its faith in the consistency of mathematics, even when Gödel's second theorem demonstrates that it cannot be proved? Do we too easily take for granted the fact that before we have even come to reason, the world is open and available to us in the first place? (Eagleton, 2009: 12)

According to John Heil (1993: 45) "Epistemological concepts of *truth, falsity* and *justification* apply primarily to beliefs and only derivatively, if at all to knowledge. Belief is thus central to epistemology." Faith is a commitment to the truth of one's beliefs.

Although faith, by definition, does not rely on evidence or proof, science and religion are by no means necessarily irreconcilable. There are those whose faith is belief *based* on testimony or authority, those whose faith is belief *guided* by testimony or authority, and those whose belief is *fixed* by testimony or authority. Reconciliation may be difficult for the first mentioned and impossible for the last, but for those for whom faith is a *guide*, there may well be tension, but conflict is not inevitable (see Paavola 2001).

Nor is the *impasse* presented by nominalism. Peirce's realism overcomes nominalism because it is temporal process and hope, the reach of subjects into the unknown—the future. "Hope," according to Eagleton, "must be fallible" (2011: 3) and "is for the most part the future tense of faith" (ibid. 68). He notes, "Peirce argues that the process of acquiring knowledge involves hope in the progress of intellectual activity itself, and that in this sense hope is one of the "indispensable requirements of logic" (ibid. 82; see CP 5:357). Knowledge, however, is never more than fact, that is, actuality. Hope reaches out to the future which is the unknown. And yet, as a line in the poem *Ailleurs, ici, partout* (Here, there, everywhere) (1946) by the French poet Paul Éluard, *L'espoir ne fait pas de poussière* (Hope raises no dust), perhaps, hope is considered too passive for modernity.

And then there is trust. The German philosopher and sociologist Georg Simmel [1858–1918] understood trust well. Simmel's champion, Guido Möllering, in his paper "The Nature of Trust: From Georg Simmel to a Theory of Expectation, Interpretation and Suspension" (2001), explains. Of particular importance from Möllering's perspective is "the recognition of affect besides reason, and system trust besides personal trust," from the perspective taken here of greatest importance is that Simmel's work is process based— trust powers process. Simmel's "leap" is reminiscent of the "leap of faith" described by Søren Kierkegaard [1813–1855]. Möllering (pp. 405–406) tells us that according to Simmel:

> To "believe in someone," without adding or even conceiving what it is that one believes about him, is to employ a very subtle and profound idiom. It expresses the feeling that there exists between our idea of a being and the being itself a definite connection and unity, a certain consistency in our conception of it, an assurance and lack of resistance in the surrender of the Ego to this conception, which may rest upon particular reasons, but is not explained by them. (Simmel, 1900: 179)

In a footnote Möllering (p. 405) brings attention to the fact that in translation of the German *Vertrauen*, "the terms confidence and trust are both used 'according to context'" (Simmel 1908, 1950: 345). He then points out that for Simmel, confidence is "an antecedent or subsequent form of knowledge" (1908, 1950: 318). Confidence, Simmel further notes, "is intermediate between knowledge and ignorance about a man [which] is a logical consequence of the view that complete knowledge or ignorance would eliminate the need for, or possibility of, trust" (1908, 1950: 318). Simmel stresses reciprocity and the relational quality of trust, whereas other sources consider trust mainly as an individual's state of mind. Ultimately, as Möllering (p. 407) notes, Simmel "attributes a high moral value to trust which makes it a rather special medium of social exchange":

> For, in the confidence of one man in another lies as high a moral value as in the fact that the trusted person shows himself worthy of it. Perhaps it is even more free and meritorious, since the trust we *receive* contains an almost compulsory power and to betray it requires thoroughly positive meanness. By contrast, confidence is 'given'; it cannot be requested in the same manner in which we are *requested* to honor it, once we are its recipient. (Möllering, 1908, 1950: 348)

Trust, as Simmel conceived it, bears an uncanny likeness to Peirce's metaphysical concept of *agape*.

Chapter 3

Belief

In the first of his 1903 Harvard Lectures, Peirce stated, "Belief consists mainly in being deliberately prepared to adopt the formula believed in as the guide to action" (PAP. 116, 1903). The response to how we know that belief is nothing but this, which he had given in his 1878 article "How to make our ideas clear," "carried this back to a psychological principle" (ibid.). Twenty-five years later he felt the need to clarify this because, as he said, "in the first place, this was not very clearly made out and in the second place, I do not think it satisfactory to reduce such fundamental things to facts of psychology" (ibid.).

Taking Peirce's second point first, made largely in response to William James's 1898 lecture. Peirce saw James as misconstruing his ideas and, feeling the need to set the record straight, declared "pragmatism is not a *Weltanschauung* but is a method of reflexion having for its purpose to render ideas clear" (CP5.13 Fn1 Para 4/6, 1907).

Regarding his first point, Peirce thought that in order to make our ideas clear, and before we are in a position to consider how to interpret propositions of belief, it is first necessary to differentiate between belief and assertion. Writing on the pragmatic maxim, he stated that

> the act of assertion is an act of a totally different nature from the act of apprehending the meaning of the proposition and we cannot expect that any analysis of what assertion is (or any analysis of what judgment or belief is, if that act is at all allied to assertion), should throw any light at all on the widely different question of what the apprehension of the meaning of a proposition is. (PAP. 116-117, 1903)

Over the years, he discussed the nature of assertion (see CP2.332-343, 1895; CP3.432-437, 1896; CP5.411-412, 1905; CP5.546-548, 1908) and from these excerpts, one can discern two relevant points.

Assertions are particular and are made in the past and stop in the present tense, which means in the main that they are bivalent and are not subject to generalization—they are nominalistic.

An example is the Pledge of Allegiance of the United States of America which is recited daily by millions of Americans. Written in 1892 by the American Baptist minister and Christian socialist Francis Bellamy for publication in the widely circulated *Youth's Companion*, it originally read:

> I pledge allegiance to my flag and the republic for which it stands, one nation, indivisible, with liberty and justice for all.

In that same year "to" was included. Bellamy had initially also considered using the words *equality* and *fraternity* but decided against it, knowing that the state superintendents of education on his committee were against equality for women and African Americans. In 1923 "my flag" was substituted with "the flag of the United States" and "of America" added a year later. It was formally adopted by US Congress as the pledge in 1942. In 1954 "under God" was formally added. It now reads:

> I pledge allegiance to the flag of the United States of America and to the republic for which it stands, one nation under God, indivisible, with liberty and justice for all.

As an assertion, it is a white, Christian, male expression of a *Weltanschauung* of the United States. While it is true *that* it is asserted, asserting it does not make it true either in *what* it asserts or *for* any given asserter. If, however, it is made the subject of inquiry, as it was prior to each of the changes made to it, or by individuals or groups, for a variety of reasons, then the reality of its form, content, justification, and validity is opened to questioning. Likewise, the *Weltanschauung* of many Christian denominations is expressed in what is called the Apostles' Creed:

I believe in God,
The Father almighty,
Creator of heaven and earth.

To proclaim, "I believe" is to make an assertion. One may truly believe, yet, unless one's assertion is propositional, and one is prepared to act in accord with that proposition, the fervency of one's believing has no bearing on the truth or otherwise of that which is said to be believed. Marge Piercy, in her poem "In the Men's Room(s)," sums it up well, if somewhat sardonically:

Now I get coarse when the abstract nouns start flashing.
I go out in the kitchen to talk cabbages and habits.
I try hard to remember to watch what people do.

Had Peirce identified the difference between belief and assertion in 1877 when "The Fixation of Belief" was first published, he may have called "beliefs" fixed by tenacity, authority, or *a priorism*—"methods" that strip beliefs of propositional value—assertions. Elizabeth Cooke puts it well:

> The method of tenacity is basically the claim that "I will not inquire." The method of authority is the method by which "We will not allow you to inquire." And the *a priori* method is that by which "you have no need to inquire: It is all self-evident." (Cooke, 2007: 34)

Those who resort to the first two "methods" may well be called dogmatic, stubborn, or even tyrannical. Unfortunately, the same is not usually said of the *a priorists* because rationalization is mistaken for rationalism. Even more unfortunate is that *a priorism* is so endemic to Western ways of thinking that it has become hegemonic. The meaning of hegemony here, adapting Bullock's definition (1988), stresses not only the political and economic control exercised by a dominant culture but its success in projecting its own particular way of seeing the world, human, and social relationships so that this is accepted as "commonsense" and part of the natural order even by those who are in fact subordinated to it. Even when raw aggression is not sanctioned as a method of imposing will, there is a myriad of culturally acceptable, even institutionally lucrative ways of achieving hegemony. Accounting for hegemony when inquiring into the spread, stymieing, or mutation of ideas and concomitant decisions and actions can reveal a whole lot about what may seem otherwise puzzling to a "purely" objective observer. This, however, is speculation; testing its validity requires a method of inquiry that does not hang itself by its own petard, that is, one that does not demand absolute certainty, absolute exactitude, or absolute universality. In other words, it requires recourse to a way of investigation that recognizes the fallibility of its method, its product, and its realism in proclaiming warrant.

Because hegemony is so censorious that it does not countenance doubt, it would be well to consider, for a moment, Peirce's conception of experience. As far as he was concerned: "(e)xperience is our only teacher" (CP5.50, 1903). It is a relationship in the form of an interaction between ego and non-ego. While ever the two appear in sympathy we can give propositional value to our assertions of belief, we have reason to trust and no *reason* to doubt. The insistence on experience makes it quite impervious to will, including a will to believe or to doubt. Only a scission in a bond between our expectations and experience—what Peirce calls surprise—can call belief into doubt and give a genuine reason for critical inquiry. In addition, as John Smith notes:

> There is a further feature of experience stressed by Peirce which is important because it is fundamental . . . and that is the tendency of experience to accumulate or fund itself and thus to become established in the pervasive form of commonsense. Commonplaces, said Peirce, are in fact "universal experiences" taken for granted as the common basis for life; he was fond of contrasting the body of large, ordinary experience which he regarded as "a valuable reservoir of truth," with special and extraordinary experience by which he meant science and the results of research into recondite matters. Funded experience shared by all sustains a culture and assumes the form of wisdom that can be passed on from one generation to another; as a consequence, pervasive experiences attains to an instinctual status which sets it off from the special and controlled experience of detail which forms the substance of science. (Smith, 1978: 94)

Although it is safe to say that it is universal that the notion of common sense is taken-for-granted sense, *what* is taken as common sense is not universal. The example of the immediate effect of touching fire, used by Peirce in speaking about indubitable beliefs, is one that can be tested empirically universally and which anyone, anywhere, learns not to test for themselves. This, however, does not cut to the core of taken-for-granted beliefs that result from, say, "up-bringing," rote, enculturation, and socialization. If, say, belief about time is considered, the argument given for the universality of common sense does not hold. The way in which different cultures, communities, and even individual people experience time is not universal. Nor is what is considered empirical. Having a reason to believe does not make beliefs true; beliefs have no *secure* foundation in fact—they are *en futuro* and are reliant on the relationship between believers and the "objects" of belief, being sustained by trust, the suspension of disbelief (a term coined in 1817 by Samuel Taylor Coleridge [1772–1834]) with every breath we take, so to speak.

In "The Fixation of Belief" Peirce argued for "the method of science" as the way "by which our beliefs may be determined" (Peirce, 1877). This could strike one as contradictory in light of the comment in his 1898 Cambridge lectures that there is "no proposition at all in science which answers to the conception of belief" (RLT.112, 1898). Unless one notes at the outset that practical affairs are the province of belief as habit and that science (including philosophy) as inquiry proceeds theoretically, one is likely to experience confusion. Unlike assertions, beliefs are propositional and therefore open to inquiry. If genuine doubt springs a leak in belief, inquiry can be activated. While it is fine, for the purpose of getting on with life, to adopt the products of inquiry at particular junctures as true, this can only be on the proviso that one accepts Peirce's doctrine of fallibilism—that in a world that is evolving, there is no guarantee of certainty. This is not the fallibilism typical of contemporary theories which, as Cooke points out, "is usually committed to mere coherence."

> Peirce's fallibilism is unique in that it rests on a commitment to a kind of vague and indeterminate correspondence with the real, but dynamic world. . . . It is not really a correspondence in the traditional sense, in the sense of static ideas mapping onto static forms. (Cooke, 2007: 4)

Rosa Mayorga's (2007) study of Peirce's "Realism," which she came to call "realicism" to distinguish it from other theories of reality, reveals that he "accepts the nominalist notion that generals, or universals, are of the nature of thought, but rejects the doctrine's claim that only individuals are real" (p. 152). In his definition of "Universal" his final word is "there is no division of logical validity into universal and particular" (CP2.371, 1902). The reality of the first universe of experience, that is, of ideas is saved through "the fact that their Being consists in mere capability of getting thought, not in anybody's actually thinking them" (Peirce, 1908). Furthermore, he rejects nominalism's denial that there is any *esse en futuro*. His is not the timeless metaphysics that follows from classic logic. Peirce's semeiotic brings into focus the temporality of the concrete life world. Floyd Merrill (2005: 8) explains that by taking the idea which is the hypothesis and interacting "in here" with its signs object "out there" the interpreter may, might, or may not find a fit. "But 'truth' is not really the goal. . . . Rather, the task at hand is to draw meaning from the signs being processed by way of interpreter-sign interaction" (ibid.).

The admission of the interpretant is an admission of fallibilism; it dashes any hope of achieving certainty. The linchpin of Peirce's realistic pragmatism, following from his metaphysics of continuity, is fallibilism.

> The principle of continuity is the idea of fallibilism objectified. For fallibilism is the doctrine that our knowledge is never absolute but always swims, as it were, in a continuum of uncertainty and of indeterminacy. Now the doctrine of continuity is that *all things* so swim in continua. (PWP.356, 1897)

Following are two examples of the interplay of belief and doubt that may also highlight the difference between assertions and belief.

Example 1

Turning first to mathematics—specifically geometry and the fifth postulate of the five postulates set forth by Euclid (365–275 BCE) as an introduction in *The Elements* to prove the rest of his theorems. These postulates are:

1. To draw a straight line from any point to any other.
2. To produce a finite straight line continuously in a straight line.
3. To describe a circle with any center and distance.
4. That all right angles are equal to each other.
5. That, if a straight line falling on two straight lines make the interior angles on the same side less than two right angles, if produced indefinitely, meet on that side on which are the angles less than the two right angles.

A postulate, though it may appear indubitable, is nonetheless a proposition. The history of doubt concerning the fifth postulate traces back more than two millennia to Posidonius of Rhodes [135–51 BCE], thence to Ptolemy [83–168] and proceeds through Proclus [412–485], Girolamo Saccheri [1667–1733], Jean d'Alembert [1717–1783], Johann Lambert [1728–1777], Adrien-Marie Legendre [1752–1833], Carl Gauss [1777–1855], Nikolai Lobachevski [1792–1856], Janos Bolyai [1802–1860], Arthur Cayley [1802–1860], Bernhard Riemann [1826–1866], and Felix Klein [1849–1925].

> Klein showed that there are three basically different types of geometry. In the Bolyai-Lobachevsky type of geometry, straight lines have two infinitely distant points. In the Riemann type of spherical geometry, lines have no (or more precisely two imaginary) infinitely distant points. Euclidean geometry is a limiting case between the two where for each line there are two coincident infinitely distant points. (O'Connor & Robertson, 1996)

To my mind, the development of topology from the late nineteenth century onward has been the greatest outcome of this freeing of geometry to move beyond the two-dimensional plane of Euclidian geometry to the three-dimensional space of non-Euclidian geometry. Topology is the study of qualitative properties of objects that are invariant under a certain kind of transformation called a continuous map. Of particular concern for transformation are convergence, connectedness, and continuity.

Example 2

My second example relates to a situation in which the identification of a change in reality caused by an absence triggered surprise. A taxi driver who, many years ago, having dropped his passenger off at a hospital's ER prepared to back out of the drive. He looked in his rearview mirror; all was clear . . . yet something was wrong. He got out of the cab to check; he found his passenger passed out in front of his back wheels. Out of sight is *not* out of mind: in the absence of seeing her in his rearview mirror as he expected, he had, nevertheless, conceived her probable trajectory and speed and concluded what he had found. Because *I* was that passenger, I am alive to relate the incident in greater detail with the help of Peirce and his three *cotary* propositions of pragmatism (pragmaticism). He called these propositions *cotary* from *cos cotis*, a whetstone, because, as he said, they "appear to me to put the edge on the maxim of pragmatism" (PAP. 241, 1903).

- First: *Nihil est in intellectu quod non prius fuerit in sensu* (Nothing is in the understanding that was not earlier in the senses) is a central tenet of the empiricism of Gassendi [1582–1655], Locke [1632–1704], and Mill [1806–1873].
- Second: "perceptual judgments contain general elements, so that universal propositions are deducible from them in the manner in which the logic of relations shows that particular propositions usually, not to say invariably, allow universal propositions to be necessarily inferred from them."
- Third: "abductive inference shades into perceptual judgment without any sharp line of demarcation between them; or, in other words, our first premises, the perceptual judgments, are to be regarded as an extreme case of abductive inferences, from which they differ in being absolutely beyond criticism."

> The abductive suggestion comes to us like a flash. It is an act of insight, although of extremely fallible insight. It is true that the different elements of the hypothesis were in our minds before; but it is the idea of putting together what we had never before dreamed of putting together which flashes the new suggestion before our contemplation. (PAP. 242, 1903)

With these three contrary propositions in mind, let me take you through the taxi driver's moments between turning to look behind him and pulling the handbrake back on. The moment of looking behind before taking his foot off the brake and on the accelerator is a moment of inquiry in the manner that Peirce called "tuism." *Tuism*—from Latin *tu*, thou—is the doctrine in ethics, which puts the emphasis on the well-being of others. In general parlance it is another name for *altruism*. Max Fisch pointed out in his introduction to the first volume of *The Writings of Charles S. Peirce* (1982: xxix) that in 1891 in the Century Dictionary, Peirce defines "tuism" as "the doctrine that all thought is addressed to a second person, or to one's future self as to a second person."

Beginning with mathematics in which the taxi driver looked for patterns in the view behind him before reversing. He saw nothing obstructing his reversal down the driveway. Thankfully he was neither myopic nor had ossified beliefs. I had stepped out of the cab and headed toward its rear intending to walk around it and across to the ER. The driver had inferred my path. The moment or two it took him to stow my fare, take the wheel, put the cab in reverse gear, put his foot on the brake, take off the handbrake, look in the rearview mirror, and turn to look behind him would have given me time to clear the rear of the taxi but not enough to reach the ER entrance door. He had seen me, but now he didn't. He did not understand. He had inferred my trajectory and that it would have taken more than six or seven seconds to complete it, given that all my movements had indicated to him that I would be walking slowly, but I was nowhere to be seen on that trajectory. Humans do not vanish into thin air; he abducted that I had to be on the ground behind him. He pulled the handbrake back on, got out of the cab, came to its rear, and found me passed out.

Here we have a description of the first principle as a method: the genera-tion or creation of hypotheses, abduction, or speculation. Deduction follows abduction and has two parts: explication and demonstration. Explication is intended to render the hypothesis as perfectly distinct as possible. Deduc-tion as explication is, like abduction, an argument. Arguments "profess to enlighten us as to the rational connections of facts or possible facts" (MS(R).142:6, 1899–1900). Deduction as demonstration is deductive *argu-mentation.* "An 'Argumentation' is an Argument proceeding upon definitely formulated premises" (Peirce, 1908). The purpose of deduction is "that of collecting consequents of the hypothesis" (Peirce, 1908).

Yet there is a tension. As he wrote in his paper "The Marriage of Religion and Science" (Peirce, 1893), "Those who are animated with the spirit of sci-ence are for hurrying forward, while those who have the interests of religion at heart are apt to press back."

In this way, science and religion become forced into hostile attitudes. Science, to specialists, may seem to have little or nothing to say that directly concerns religion; but it certainly encourages a philosophy which, if in no other respect, is at any rate opposed to the prevalent tendency of religion, in being animated by a progressive spirit. There arises, too, a tendency to pooh-pooh at things unseen. (SW. 351-2. 1893)

In his remarkable paper "A Neglected Argument for the Reality of God" (Peirce, 1908) Peirce advanced his hypothesis of God. His concept of reality being the antithesis of nominalism, his argument did not pretend to be a *proof* of the *existence* of God but rather the hypothesis of God flowing from the form of abduction which he called "Musement" or Pure Play.

Play has no rules, except this very law of liberty. . . . It has no purpose, unless recreation. The particular occupation I mean . . . may take either the form of aesthetic contemplation, or that of distant castle-building . . . or that of consider-ing some wonder in one of the Universes, or some connection between two of the three, with speculation concerning its cause. It is this last kind—I will call it "Musement" on the whole—that I particularly recommend, because it will in time flower into the N.A. (Peirce, 1908)

His argument is, in fact, a nest of three arguments—the humble argument (HA), the neglected argument (NA), and the scientific argument (SA)—which are interdependent. Douglas explains:

> The HA relates to Feeling, the NA to Willing and the SA to Thinking. And these relations are relations of appeal. That is, the HA comes to us by feeling—it appeals to our feeling; the NA appeals to our need to act in the world, to be "willing"; and the SA appeals to our critical thinking, our pursuit of truth. (Anderson, 1995: 352)

For the living, feeling is indubitable. To speak of feeling as an argument, as with the HA, is to speak of feeling as reasonable, that is, controllable. One method of "control" is to engage in musement or as I prefer to call it, wonderment—there is nothing like a bit of awe for relief of the bondage of self. The humility it can engender, if one can enter into it as recommended, is not only a superior prophylactic, but it is the ground for the open-mindedness required to engage with reality. As Peirce put it, "It is simply the natural precipitate of meditation upon the origin of the Three Universes" (CP6.487, 1910) and is the source of abduction.

K. T. Fann (1970: 5) points out, "Peirce's theory of abduction is concerned with the reasoning which starts from data and moves towards hypotheses." The data here are impressions observed in any of Peirce's three universes. Observation without purpose, partiality, or breach of continuity passes into musement from which the hypothesis of God can flow. As Peirce saw it:

> The hypothesis of God is a peculiar one, in that it supposes an infinitely incomprehensible object, although every hypothesis, as such, supposes its object to be truly conceived in the hypothesis. This leaves the hypothesis but one way of understanding itself; namely, as vague yet as true so far as it is definite and as continually tending to define itself more and more and without limit. The hypothesis, being thus itself inevitably subject to the law of growth, appears in its vagueness to represent God as so, albeit this is directly contradicted in the hypothesis from its very first phase. But this apparent attribution of growth to God, since it is ineradicable from the hypothesis, cannot, according to the hypothesis, be flatly false. Its implications concerning the universes will be maintained in the hypothesis, while its implications concerning God will be partly disavowed and yet held to be less false than their denial would be. Thus the hypothesis will lead to our thinking of features of each universe as purposed, and this will stand or fall with the hypothesis. Yet a purpose essentially involves growth and so cannot be attributed to God. Still it will, according to the hypothesis, be less false to speak so than to represent God as purposeless. (Peirce, 1908)

Whole books, PhD theses, and a great number of journal articles have been written on Peirce's concept of abduction, yet the inadequate and misleading definition of "inference to the best explanation" persists. This definition is reflected in Peirce's early exploration of its meaning, shown by Anderson in his paper "The Evolution of Peirce's Concept of Abduction":

> In the 1878 view of abduction we can see the relation to Aristotle. Its particular form was that of the acceptance of a minor premiss as a hypothesis on the strength of its "fittingness" to a known premiss and a factual conclusion. Thus, if a deduction were to take the following form:
>
> Rule—All the beans from this bag are white.
> Case—These beans are from this bag.
> ∴ Result—These beans are white.
>
> Then its corresponding abduction would be:
>
> Rule—All the beans from this bag are white.
> Result—These beans are white.
> ∴ Case—These beans are from this bag. (Anderson, 1986: 148)

Peirce says that in this case of "making a hypothesis . . . I at once infer as a probability, or as a fair guess, that this handful was taken out of that bag" (CP2.623, 1893).

Further comparison with the two more familiar forms of reasoning can serve for those who want an explanation in bytes. "*Abduction* is the process of both generating hypotheses and selecting some for further pursuit, *deduction* draws out their testable consequences, while *induction* evaluates them" (CP5.171, 1903; CP6.468-477, 1908). Abduction is reasoning *toward* a hypothesis while deduction and induction are reasoning *from* a hypothesis. As Peirce puts it:

> Deduction must include every attempt at mathematical demonstration, whether it relates to single occurrences or to "probabilities"—that is, to statistical ratios; Induction must mean the operation that induces an assent, with or without quantitative modification, to a proposition already put forward, this assent or modified assent being regarded as the provisional result of a method that must ultimately bring the truth to light; while Abduction must cover all the operations by which theories and conceptions are engendered. (CP5.590, 1903)

On a scale of low to high: abduction is low in security (approach to certainty) but high in what Peirce called uberty (rich growth, fertility, abundance); induction is the reverse.

In his paper "The Logic of Drawing History from Ancient Documents" (1901) Peirce adopted Aristotle's term *apagögé*, which he translated as "abduction" (see CP7.249, 1901). Though he retained the word, his logic went beyond the syllogistic method of Aristotle and in accord with this, his understanding of *apagögé* expanded. Anderson points out (1986: 147) "in working through Aristotle's discussion of *apagögé*, Peirce lays the foundations for a type of reasoning which has a logical form but which is also a lived process of thought." His mature meaning of abduction retained an element of inferencing but became more of an insight, that is, a creation of the mind. It is not about "probability" which is the output of scientific reasoning but rather about reasonable "may-be," on which scientific reasoning depends. Karl-Otto Apel argues that for Peirce

> cognition was seen as a historical process manifested in language and society which, from the standpoint of its unconscious foundations, forms a continuum with the evolutionary process of nature, but which, at the other conscious extreme, is subject to "self-control" through normative logic. (Apel, 1967: 160)

More than anything, it is this normative science of reasoning that distinguishes the psychological leanings of pragmatism as espoused by William James in 1898 from that which Peirce developed an apology after that date and called pragmaticism. From the perspective of pragmaticism, "The hypothesis of God's Reality . . . is connected so with a theory of the nature of thinking that if this be proved so is that" (CP6.491, 1910); the HA, through the NA, is open to the "SA's primary appeal is to the category of thinking—to a critical pursuit of the truth" (Anderson, 1995: 356).

Unfortunately, no English dictionary that I found gives Peirce's meaning of abduction. The SOED tells us that "abduction" *noun* is from the Latin *ab* off, away, from + *duct-* pa. ppl. stem of *ducere* to lead + *ion* forming noun denoting verbal action or an instance of it or a resulting state or thing. In English, from *E17* it had the general meaning of "a leading away." From the *M17* it meant in anatomy "movement of a limb etc. outward from the median line." Supposedly from *L17* it had taken the meaning in logic as "a syllogistic argument with the major premiss certain, the minor only probable." No reference to this use of abduction is given, so I can only presume "*17*" is a typo. The fourth and most commonly understood meaning, dating to *M18*, is the act of illegally carrying off or leading away a person. Wondering from whence "illegally," I discovered that its etymology is from Latin *abductiō* robbing + *n* forming a perfect passive participle to form a noun of action.

In the spirit of Peirce's developed concept of abduction, I like the meaning given in anatomy, which I picture as an opening out from central control which is both brain and mind. Consider that dance of breathtaking beauty the flamenco. Flamenco—from the Spanish *flama* fire or flame + *enco* having a quality of—is, as *flamenco puro*, improvized rather than choreographed and, unlike many traditional dances, has no purpose beyond itself.

In his musement on his hypothesis Peirce (perhaps unknowingly) makes use of Bayes theorem which proposes that evidence confirms the likelihood of a hypothesis only to the degree that the appearance of this evidence would be more probable with the assumption of the hypothesis than without it. Peirce points out, "every hypothesis, as such, supposes its object to be truly conceived in the hypothesis" (Peirce, 1908). He also brings to our attention that "it is a chief function of an explanatory hypothesis . . . to excite a clear image in the mind by means of which experiential consequences of ascertainable conditions may be predicted" (CP6.489, 1910).

Gathering evidence begins induction, the process "of ascertaining how far those consequents accord with Experience and of judging accordingly whether the hypothesis is sensibly correct, or requires some inessential modification, or must be entirely rejected" (Peirce, 1908). Induction is a three-staged process, the first, which is Classification, is the process "by which general Ideas are attached to objects of Experience" (ibid.). The second stage "the testing argumentations" called by Peirce the "Probations" are of two kinds. The first, Crude Induction, "is the weakest of arguments, being liable to be demolished in a moment. . . . The other kind is Gradual Induction, which makes a new estimate of the proportion of truth in the hypothesis with every new instance" (ibid.) and "is either Qualitative or Quantitative" (ibid.). The third stage, called the "Sentential part" by Peirce, "appraises the different Probations singly, then their combinations, then makes self-appraisal of these very appraisals themselves, and passes final judgment on the whole result" (ibid.).

A "sententia" is a maxim. The maxim of appraisal Peirce would be referring to his "pragmatic maxim" was first put forward by him in his paper, "How to Make Our Ideas Clear" (1878):

> Consider what effects, which might conceivably have practical bearings, we conceive the object of our conception to have. Then, our conception of these effects is the whole of our conception of the object. (Peirce, 1878)

He later revisited and confirmed this:

> The doctrine that the whole meaning of a conception expresses itself in practical consequences, consequences either in the shape of conduct to be recommended, or in that of experiences to be expected, if the conception be true; which consequences would be different if it were untrue and must be different from the consequences by which the meaning of other conceptions is in turn expressed. (CP5.2, 1902)

But what to believe.

The skeptic would have it that you believe nothing. Pyrrho of Elis (360-270 BCE) originated classical skepticism. Since there are plausible arguments for both sides of any issue, Pyrrho argued, the only rational practice is to suspend all judgments, abandon worries of every kind, and live comfortably in an appreciation of the appearances. For the Pyrrhonians, given that neither the sense impressions nor the intellect, nor both combined, is a sufficient means of knowing and conveying truth, one suspends judgment on dogmatic beliefs or anything non-evident.

Agrippa the skeptic (first century) is regarded as the author of "five grounds of doubt" or tropes, which are purported to establish the impossibility of certain knowledge. The tropes are:

1. Dissent: The uncertainty of the rules of common life and of the opinions of philosophers.
2. Progress *ad infinitum*: All proof requires some further proof and so on to infinity.
3. Relation: All things are changed as their relations become changed, or, as we look upon them from different points of view.
4. Assumption: The truth asserted is merely a hypothesis.
5. Circularity: The truth asserted involves a vicious circle.

Sextus Empiricus (160–210) defended the practical viability of Pyrrhonism as the only way of life that results in genuine *ataraxia* (ἀταραξία "tranquility"). In his medical work, tradition maintains that he belonged to the "empiric school," as reflected by his name.

If the purpose of adopting skepticism was *ataraxia*, then the Epicureans qualify. For them *ataraxia* was synonymous with the only true happiness possible for a person. It signifies the state of robust tranquility derived from eschewing faith in an afterlife, not fearing the gods because they are distant and unconcerned with us, avoiding politics and vexatious people, surrounding oneself with trustworthy and affectionate friends and, most importantly, being an affectionate, virtuous person, worthy of trust.

The philosophy of Epicurus (341–270 BCE) was a complete and interdependent system, involving a view of the goal of human life (happiness, resulting from the absence of physical pain and mental disturbance), an empiricist theory of knowledge (sensations, including the perception of pleasure and pain, are infallible criteria), a description of nature based on atomistic materialism and a naturalistic account of evolution, from the formation of the world to the emergence of human societies.

The Stoics, too, sought mental tranquility seeing *ataraxia* as something to be highly desired and often made use of the term, but for them the analogous state attained by the Stoic sage was *apatheia* (ἀπάθεια; from a-"without" and pathos "suffering" or "passion"). *Apatheia* in Stoic philosophy refers to a state of mind where one is not disturbed by the passions. It may be best translated by the word "equanimity" rather than "indifference." The word *apatheia* has a quite different meaning to the modern English "apathy," which has a negative connotation. It would appear to translate to the modern notion of objectivity.

Neither the Epicureans nor the Stoics were the first nor were they the last to believe these apposing positions to be the way of truth. Around two centuries earlier, empiricist Heraclitus (535–475 BCE) was proclaiming that you can't step in the same river twice, while Parmenides (515–450 BCE) firmly believed that nothing changes.

Parmenides, the founder of the Eleatic School and of Western metaphysics, was a monist. He argued that "being" is an indivisible "whole" and that there is no epistemological validity to sense experience such as that expounded by Heraclites, rather logical standards are the criteria of truth. A couple of ways around these dilemmas were the Pluralism of the likes of Anaxagoras (500–428 BCE) and Atomism as theorized by Leucippus (*fl.* 440 BCE) and systematized by Democritus (460–457 BCE).

What thought of the big names informing Western thinking, Socrates (469–399 BCE), Plato (427–347 BCE), and Aristotle (384–322 BCE)? Socrates questioned Athenians about their moral, political, and religious beliefs. His questioning technique, called dialectic, has greatly influenced Western philosophy. He is alleged to have said, "The unexamined life is not worth living." His student Plato was the innovator of the written dialogue and dialectic forms in philosophy. Plato is also considered the founder of Western political philosophy. His most famous contribution is the theory of Forms known by pure reason, in which Plato presents a solution to the problem of universals known as Platonism (also ambiguously called either Platonic realism or Platonic idealism). Aristotle was the first to pay serious attention to the problem presented by discreteness and continuity to develop a universal method of reasoning by means of which it would be possible to learn everything there is to know about reality. The aim of his logical treatises was to develop a universal method of reasoning. His arguments were rejected by the Epicureans but upheld by the Stoics. Likewise, the concept of infinitesimals, first raised by Democritus, was controversial and remains largely contested into the present.

More than 2.5 millennia since the ancients, we remain unable to guarantee certainty; we can believe with all sincerity but never with a warrant. Whatever our position we all have still minimally to get by. Nevertheless, if our taken-for-granted beliefs are taken by us to be indubitable, we are resistant to learning and to inquiry; to thinking in ways that contest the truth of our beliefs. We may be knowledgeable, but we are bereft of broad understanding. Belief may well sustain us, but as Sportin' Life sang in George Gershwin's *Porgy and Bess*, we need to understand "it ain't necessarily so."

While Peirce accepted that all reality is not known, such acceptance is not tantamount to conceding to incognizability. He did not countenance the inexplicable, seeing such an idea, exemplified, for instance, in Kant's "thing-in-itself," as incoherent. He opted for fallibilism. Cooke points out that "the commitment to fallibilism is a commitment to the belief that there are knowable truths about which inquirers are fallible" (Cooke, 2007: 23). She speaks of Peirce's model of knowledge as "adaptionalist" noting that on his account, "knowledge is not 'true belief' but is the integration of rational habits with ontological habits" (p. 28). What is interesting in Peirce's philosophy is that "both certainty and skepticism are seen as posing roadblocks to inquiry instead of being virtues or goals of inquiry" (p. 32).

In his 1899 paper "First Rule of Logic" Peirce presents fallibilism as a propositional attitude:

> Upon this first and in one sense this sole, rule of reason, that in order to learn you must desire to learn and in so desiring not be satisfied with what you already incline to think, there follows one corollary which itself deserves to be inscribed upon every wall of the city of philosophy:
> Do not block the way of inquiry.
> Although it is better to be methodical in our investigations and to consider the economics of research, yet there is no positive sin against logic in trying any theory which may come into our heads, so long as it is adopted in such a sense as to permit the investigation to go on unimpeded and undiscouraged. On the other hand, to set up a philosophy which barricades the road of further advance toward the truth is the one unpardonable offence in reasoning, as it is also the one to which metaphysicians have in all ages shown themselves the most addicted.
> Let me call your attention to four familiar shapes in which this venomous error assails our knowledge:
> The first is the shape of absolute assertion.
> The second bar which philosophers often set up across the roadway of inquiry lies in maintaining that this, that and the other never can be known.
> The third philosophical stratagem for cutting off inquiry consists in maintaining that this, that, or the other element of science is basic, ultimate, independent of aught else and utterly inexplicable—not so much from any defect in our knowing as because there is nothing beneath it to know.
> The last philosophical obstacle to the advance of knowledge . . . is the holding that this or that law or truth has found its last and perfect formulation. (RLT.178-180, 1899)

Knowing is static, referring to discrete facts organized supportively; understanding is active, working facts in context to fit a cognizable metaphysics. Achieving understanding rather than truth is the aim of inquiry. Understanding comes with having a clear idea of meaning and that's where Peirce's method comes right to the fore. With information-qualified interaction, we can come to a clearer understanding of the breadth and depth of "mattering."

Chapter 4

Information

Anthony Kenny, in the "General Introduction" to his *A New History of Western Philosophy* (2010), tells his readers:

> Philosophy is not a matter of expanding knowledge or acquiring new truths about the world; the philosopher is not in possession of information that is denied to others. Philosophy is not a matter of knowledge; it is a matter of understanding, that is to say, of organizing what is known. (Kenny, 2010)

He has equated information with knowledge—with content—and differentiated it from understanding which he equates with organizing content—with process. Unfortunately, Kenny does not appear to realize that, conceived of as content, information is not knowledge, nor, as a process, is understanding the equivalent of organizing. No amount of progress in organizing content, that is, knowledge, will bring information to light. Nor will it produce the kind of fit that is understanding. To give him his due, he was expanding on his explanation of progress in philosophy. For him:

> The difference between what we might call the Aristotelian and the Wittgensteinian attitude to progress in philosophy derives from two different views of philosophy itself. Philosophy may be viewed as a science, on the one hand, or as an art, on the other. (Ibid.)

He argues that philosophy is neither art nor science but rather lies somewhere between the two. This, I agree, is more appropriate than the traditional view of philosophy as art, or of Peirce's view that, as engaged in discovery through inquiry, it is science. To clarify this somewhat:

During 1855, the year Peirce entered college and began the study of Kant, he and his then best friend, Horatio Paine, spent hours discussing Friedrich

Schiller's *On the Aesthetic Education of Man in a Series of Letters*. I too read this to see what captivated the two boys. Of particular interest to my mind is the dilemma and its horns outlined in the first footnote of his Thirteenth Letter:

> Once we assert the primary, and therefore necessary, antagonism of the two impulses, there is really no other means of preserving the unity in Man except by the unconditional *subordination* of the sensuous impulse to the rational. But the only result of that is mere uniformity, not harmony, and Man remains forever divided. Subordination there must indeed be, but it must be reciprocal; for although limits can never establish the Absolute—that is, freedom can never be dependent on time—it is equally certain that the Absolute by itself can never establish the limits, that conditions in time cannot be dependent on freedom. Both principles are at once mutually subordinated and coordinated—that is, they act and react upon each other; without form no matter, without matter no form. . . . Necessary as it may be that feeling should decide nothing in the realm of reason, it is equally necessary that reason should not presume to decide anything in the realm of feeling. In the very act of awarding to either of them its own territory we are shutting the other out, and giving each of them a boundary which can be crossed only to the injury of both. (Schiller, 1795, 1801, 1954: 68)

Recalling this time with his friend forty-something years later, Peirce came to the realization that esthetics (aesthetics) informs ethics. This is the connection between philosophy and science that Peter Forrest (1991) calls "aesthetic understanding." Once one grasps Peirce's method, one sees that it is as creative as any work of art not least because of his engagement of imagination through the process of abduction. But so too is science through its ability to reveal the reality of the evolving and developing cosmos. It is art-semeiotic-science.

I have focused on Peirce's third heuretic science—Formal Science: "The means and the method for conducting inquiry" (MS 108, begun in 1865; W1: 303)—particularly on his philosophy, although I was also concerned with illuminating his idea of pure mathematics as the genesis of creativity. At the center of his philosophy are his normative sciences. If the fractal pattern of his architectonic in relation to his categories maintains, then for his philosophy: his phenomenology is a First of a Second: his normative sciences, of a Third: his metaphysics. In like manner for his normative sciences, aesthetics is a First of a Second—ethics—and of a Third—logic. Peirce's first heuretic science—positive science—"is experimental science . . . conducted for the purpose of determining the ontological validity of a hypothesis" (ibid.). His second heuretic science, "semeiotic—the science of representations . . . is the process of reasoning . . . conducted for the purpose of determining meaning. It entails creation, explication and verification of a hypothesis" (ibid.).

Peirce outlined a theory of information a combination of his three heuretic sciences—which he defined as "the connection of form with matter" (CP2.418n, 1893) I=MF. This he explained more fully in his presentation to the American Academy, November 13, 1867, "Upon logical comprehension and extension" (W2.70-86) together with some footnotes he added in 1893 (CP2.407: Fn P1 & CP2.422: Fn P1). His theory of information provides the resolution to the either/or problem of empirical OR rationalist approaches to inquiry. His identification of abduction as the creative ground of mathematics, phenomenology as the clarifier of observation, and rhetoric as interpretive boost of logic revealed the gateway to his reconstructed metaphysics and ultimately the means of discovering the information of reality. André de Tienne (2006) points out that "Peirce's pragmatic theory of information is indissolubly connected to his semeiotic theory of propositions, itself an elaborate outgrowth of the traditional subject-predicate propositional logic and of the logic of relatives." The theoretical physicist Frank Wilczek in speaking of the behavior of protons and in particular "the indeterminism for which quantum mechanics is famous and which caused Einstein such anguish" has this to say:

> This abundance of coexisting possibilities in the phenomena and in the quantum theory that describes them, defies traditional logic. The success of quantum theory in describing reality transcends and in a sense unseats classical logic, which depends on one thing being "true" and its contraries "false." (Wilczek, 2008: 46)

Notwithstanding that logic is concerned with truth, it is truth *en futuro*. Logic, for Peirce, is a way of dealing with ontology and is directed at the elucidation of reality. As de Tienne (2006) sees it: "truth is not a disembodied property of arbitrary definitions, but a consequential measure of a proposition's capacity to represent 'real things,' that is, things rooted in an actual world of action and reaction." In a footnote to this he further points out that Peirce explained (EP2: 278, 1903) that "nominal definitions are propositions in the imperative mood and thus not real propositions, which require the indicative mood: a real anchor is a *sine qua not*." Andrew Smith (2009) argues that Peirce offered an account of the *pragmatic* meaningfulness of truth: "Of what the experience of grasping truth would involve" (ibid.). In this light, while propositions may be *linguistically* meaningful, they are pragmatically meaningless if they cannot be explained in terms of observable effects associated with experiencing their object.

> Reality, according to Peirce, is the object of true propositions. And true propositions, which are associated with true beliefs, remain indefeasibly settled not because inquirers are fated to accept them but because they stand up effectively to whatever scrutiny they face. (Smith, 2009)

Peirce conceded to bivalence only in the case of a dyadic value system:

> The simplest of value systems serves as the foundation for mathematics and, indeed, for all reasoning, because the purpose of reasoning is to establish the truth or falsity of our beliefs and the relationship between truth and falsity is precisely that of a dyadic value system. (MS.6)

Mathematics reveals necessary but *hypothetical* truth. Positive facts are dyadic—they are secondness—but to be known, they must be triadic, that is, they must be meaningful. In situations of what Peirce called "buried secrets"— past facts that may be irretrievable—if they lack practical significance in accordance with the pragmatic maxim, truth-value is of no conceivable consequence.

Through inquiry, information emerges by way of a continuous semeiotic process. Knowledge is discovered information; it is only ever partial and never certain. Furthermore, as Peirce explained in a letter to William James (EP 2:495, 1909) in speaking of what he called the Immediate Object and the Dynamical Object, the Immediate Object is "the Object as it is regardless of any particular aspect of it, the Object in such relations as unlimited and final study would show it to be" whereas the Dynamical Object "is the Object that Dynamical Science (or what is called 'Objective' science) can investigate."

The Dynamical Object, as experience is indubitable, of itself it is not intelligible. It is the Immediate Object that is cognized through a continuous semeiotic process. Peirce's semeiotic, or system of signs, is implied by his pragmatic maxim. As Hausman notes:

> if meaning consists of ever-widening connected consequences, then the pragmaticist's maxim concerns a dynamic system of references—a growing web of consequences and their interpretations, which not only refer to but which themselves also refer to further interpretations. (Hausman, 1993: 57–58)

In this regard, Peirce distinguished between the Immediate Interpretant, the Dynamical Interpretant, and the Final Interpretant:

> In the first place, the Immediate Interpretant, which is the interpretant as it is revealed in the right understanding of the Sign itself and is ordinarily called the *meaning* of the sign; while in the second place, we have to take note of the Dynamical Interpretant which is the actual effect which the Sign, as a Sign, really determines. Finally there is what I provisionally term the Final Interpretant, which refers to the manner in which the Sign *tends* to represent itself to be related to its Object. (CP4.536, 1906)

Informed meaning of dynamic objects is the purpose of inquiry. As events, dynamical objects are always evolving and thus, in a manner of speaking, always outstrip any knowledge gained. As stated earlier, Peirce's pragmatism is a method of determining meaning not a doctrine of the truth of things (MS 322, 1907). His logic or semeiotic is not, therefore, equivalent to classical logic. In speaking of his pragmatic elucidation of truth Andrew Smith (2009) explains that for Peirce, truth is an epistemic ideal. Propositions can be said to be true and continue to be true and to the extent they stand up effectively to whatever scrutiny they face. Any scrutiny, perforce, involves engagement in semeiotic.

The difference between information and knowledge in Peircean terms is the difference between the "Immediate Object" and the "Dynamical Object," where "object" may be better understood as "event." Knowledge is that which is found by collateral experience, that is, it is what is discovered by means of the heuretic sciences and which can be learnt; it is the provenance of discoverers and learners. Knowledge as the difference between the "Immediate Object" and the "Dynamical Object" is meant literally: it is that which might be, is, or may be *taken* from the latter as represented by the former. Realizing this difference also suggests why Peirce's focus was on meaning rather than truth and on fallibilism rather than certainty. The need is to consider semeiotic not as language for communicating but as signs that inform. Seen this way, it is as relevant to the physical sciences as it is to the life sciences of the creating of something from nothing—of "mattering."

According to Peirce "continuity is the leading conception of science" (CP1.62, 1896). He describes continuity as "merely a discontinuous series with additional possibilities" (CP1.170, 1897). These additional possibilities are infinitesimals. Peirce says of continuity:

> How can one mind act upon another mind? How can one particle of matter act upon another at a distance from it? The nominalists tell us this is an ultimate fact —it cannot be explained. Now, if this were meant in (a) merely practical sense, if it were only meant that we know that one thing does act on another but that how it takes place we cannot very well tell, up to date, I should have nothing to say, except to applaud the moderation and good logic of the statement. But this is not what is meant; what is meant is that we come up, bump against actions absolutely unintelligible and inexplicable, where human inquiries have to stop. Now that is a mere *theory* and nothing can justify a theory except its explaining observed facts. It is a poor kind of theory which in place of performing this, the sole legitimate function of a theory, merely supposes the facts to be inexplicable. It is one of the peculiarities of nominalism that it is continually supposing things to be absolutely inexplicable. That blocks the road of inquiry. But if we adopt the theory of continuity we escape this illogical situation. We may then say that one portion of mind acts upon another, because it is in a measure immediately present to that other; just as we suppose that the infinitesimally past is in a measure present. And in like manner we may suppose that

one portion of matter acts upon another because it is in a measure in the same place. (PWP.355-6, 1897)

Most of the contemporary material about information is concerned with information technology and artificial intelligence. The book of papers edited by Paul Davies and Niels Henrik Gregersen, *Information and the Nature of Reality: From Physics to Metaphysics* (2010), is different. If several of the authors are courageous enough to up their ability to "accommodate"—that is, change their mental schemata—rather than try to "assimilate"—squeeze objective reality—to fit their idea of it, they can progress their respective disciplines. In their introduction to the book (which I will consider in some detail because of its change of focus regarding information) the editors comment that

> information makes a causal difference to our world—something that is immediately obvious when we think of human agency. But even at the quantum level, information matters. (Davies & Gregersen, 2010: 7)

Frank Wilczek, in his *The Lightness of Being: Mass, Ether and the Unification of Forces* (2008), entitles one of the chapters "The Bits within the Its" and speaks of *embodied ideas*; Paul Davies in his paper "Universe from Bit" (pp. 65-91) has the heading "It from bit." Both in their way are talking about information. Davies writes about the laws of physics *being* informational statements. If, when considering, say, standard thermodynamics rather than describing entropy as a measure of the degree of disorder of a system, one substitutes it for the terminology of Leo Szilard (1929, 1964) then "entropy can be regarded as a measure of ignorance."

> Ignorance is the flip side of information, so we may deduce a mathematical relationship between entropy and information. . . . The link between information (loss) and area seems to be a very deep property of the universe . . . a so-called *holographic principle*. (Davies, 2010: 78–79)

Furthermore, to understand Peirce's metaphysics of developmental evolution is to realize that Davies's statement that "the laws of physics are inherent in and emergent with the universe, not transcendent of it" (p. 83) assumes that the universe is imbued with information.

Jesper Hoffmeyer, associated with the International Society of Biosemeiotic Studies, in his paper "Semeiotic Freedom: An Emerging Force" (pp. 185-204) demonstrates an understanding of Peirce's "semeiotic," unlike Terrence Deacon (pp. 146–169) who relies on Charles Morris's "semiotics" (see glossary for difference). Hoffmeyer introduces the subject of "information in evolution" noting that "something is added to our understanding when we talk about information rather than just about chemistry" (p. 187). He suggests, "The heuristic value of the information concept is connected to the role that history (evolution) plays in the life of cell and organisms" (ibid.).

To speak of living systems as semiotic systems in their own right is, I take it, saying that a semiotic system does not stand for something else but rather that *it is*. While semiosis may be instrumental, that is, it may be used epistemologically by (among others) biosemeioticians; biosemeioticians and other subjects (and predicates) are ontological. Hoffmeyer notes that

> signs, however, are not causes in the traditional sense of (Aristotelian) efficient causality, for the effect of a sign is not compulsory but depends upon a process of interpretation and the interpretation may well be—and probably most often is—mistaken. (Hoffmeyer, 2010: 190)

> The historical nature of the world has profound consequences for the study of life. . . . The principle of natural selection, of course, greatly helps us in explaining the widespread adaptedness of biosystems, but we need an additional principle that would solve the fundamental question of the "aboutness" of life, the never-ending chain of attempts by living systems to come to terms with their conditions of life. (Ibid. 192)

He then points out that "'aboutness' ... is not derivable from the principle of natural selection for the simple reason that it is required for natural selection to operate in the first place" (p. 192). As he sees it:

> The difficult problem to solve in any theory of the origin of agency ("aboutness") and life is how to unify two normally quite separate kinds of dynamics: a dynamics of chemical interaction patterns and a dynamics of signification or semiosis. This immediately places this question in the contextual situation of the environment. (Ibid. 193)

Hoffmeyer then shows the operation of semeiotic freedom.

> In the semeiotic understanding . . . the chemotactic machinery serves to integrate the sensing of the outer world to the reality of the inner world as this reality is described in the self-referential, or generic, systems. (Ibid. 195)

He follows this by declaring that "allowing for semeiotic freedom in the organic world significantly changes the task of explaining emergent evolution, because semeiotic freedom has a self-amplifying dynamic" (p. 196). As he puts it: "Instead of the Cartesian either-or thinking, biosemeiotics institutes a more-or-less thinking" (p. 197).

He explains this semeiotic emergence through the operation of downward causation, which he says "may be seen as an attempt to express parts of what used to be called 'final causation:'" (p. 197)—what Peirce called the finious

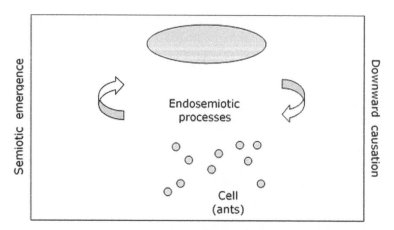

Figure 4.1 Downward causation operates through indexical sign relations, that is, the values of system parameters and interpreted by lower-level agents as indexical signs. But this state of affairs in itself presupposes the formation in the first place of a large-scale pattern with a behavior that stabilizes the semeiotic interaction between parts. *Source*: Adaptation of Hoffmeyer (2010: 200).

process of causation. For Hoffmeyer "semeiotic emergence and downward causation are two sides ... of the same coin, which I would prefer to call 'semeiotic causation,' bringing about effects through a process of interpretation" (p. 98). The diagram below illustrates how this works. This particular example also illustrates how Peirce's notion of community *vis-á-vis* individuals operates in a system as shown in Figure 4.1.

Hoffmeyer concludes, under the heading of "Biosemeiotics and God," that

> the biosemeiotic approach ... overturns the need for, or legitimacy of, the argument for intelligent design. Biosemeiotics does not logically entail any stance on the presence or absence of a transcendental creator. (Ibid. 202)

This position not only does away with the need to affirm or deny a belief in God in science but that it does so without the need to side-step the issue in the pursuit of discovery in the natural sciences. Peirce railed against dogma—against "blocking the road of inquiry"—not against belief (or non-belief) in God. The solution to the matter of natural intentionality, which attention to Peirce's semeiotic successfully addresses, is one that can serve the physical sciences by following the lead of biosemeioticians such as Hoffmeyer.

Claus Emmeche, of the International Society of Biosemeiotic Studies, says, "Understanding C. S. Peirce is mandatory in biosemeiotics." He shows just such an understanding when he and coauthors Charbel Niño El-Hani and João Queiroz in *Genes, Information and Semiosis* (2009) point out that it has been shown that "there are several epistemological flaws in the treatment of 'biological information' in 'systems biology,' most importantly, a troublesome confusion between information handled by organisms and information handled by the observer." Hoffmeyer and Emmeche had previously argued that "both the metaphors of 'information' and 'program' make the dilemmas of form and substance disappear by simply treating DNA (a substance) and program (information/or potential form) as one and the same thing" (2009: 18). This "one and the same thing" is information as Peirce formulated it: I = MF. Using Peircean terms to build a framework for theorizing information as a process, they formulate "(Information = semiosis) A triadic-dependent process through which a form embodied in the Object in a regular way is communicated to an Interpretant through the mediation of a Sign" (p. 96). Their definition of process follows that of Nicholas Rescher in *Process Metaphysics: An Introduction to Process Philosophy* (1996: 38): "A coordinated group of changes in the complexion of reality, an organized family of occurrences that are systematically linked to one another either causally or functionally." "The exploration of the scientifically new continent of 'meaning' has just begun" is how Liz Else describes biosemiotics in "A Meadowful of Meaning" (*New Scientist*: August 21, 2010, 28–31).

Information as the product of matter and form through time is complex and is in need of a theory to grasp its meaning. Some of the developments of such complexity theories have been mapped by Brian Castellani (2013, http://sci-maps.org/mapdetail/map_of_complexity_sc_154). His was one of ten maps exhibited by the Places & Spaces: Mapping Science Project in their 9th iterations (2013) *Science Maps Showing Trends and Dynamics* and is intended to show complexity science's major intellectual traditions, leading scholarly themes and methods, as well as key scholars who founded or pioneered work. While information is not the focus of the map, scholars who were concerned with information can be identified in each of the five trajectories. Gregory Bateson [1904–1980], for example, in *Form, Substance and Difference, in Steps to an Ecology of Mind* (1972) spoke of "information" as a "difference that makes a difference."

Claude Shannon and Andrey Kolmogorov are identified as main players by Pieter Adriaans and Johan van Benthem, the editors of *Philosophy of Information*, in their "Introduction: Information Is What Information Does," when describing three stances from which to consider information:

Information-A Knowledge, logic, what is conveyed in informative answers
Information-B Probabilistic, information-theoretic, measured quantitatively
Information-C Algorithmic, code compression, measured quantitatively. (Shannon & Kolmogorov, 2008: 13)

Oversimplifying a bit, *A* is the world of epistemic logic and linguistic semantics, *B* that of Shannon information theory, linked to entropy in physics, and *C* that of Kolmogorov complexity, linked to the foundations of computation. We do not feel that these are opposing camps but rather natural clusters of themes and research styles. Thus, we felt that all of these need to be represented in our Handbook since only their encounter gives us the proper canvas for philosophical inquiry.

Nevertheless, Casagrande, along with Castellani and the contributors to *Philosophy of Information* (2008), focus on epistemological rather than ontological issues of metaphysics. However, their identification of process does warrant consideration when attempting to understand reality: cybernetics, globalization, self-organization including self-organized criticality, fractal geometry, autopoiesis, adaption, the "butterfly" effect, nonlinear dynamics, synergetics, swarm intelligence, emergence, and the list goes on.

David Casagrande's "Information As Verb: Re-conceptualizing Information for Cognitive and Ecological Models" (1999: 4) posited that

> current notions of information are inadequate for ecological and cognitive models because they 1) only account for information gain that results from reducing uncertainty; 2) assume binary logic; 3) fail to account for semantics and pragmatics; and 4) cannot account for shared and externalized cognition.

He proposed an alternative model "which treats information as a process of state change." The physicist and writer Mark Buchanan's *Ubiquity: The Science of History* (2000) does the same for physics. He explains for the non-physicist, the non-equilibrium phenomena of criticality. The theoretical physicists Per Bak and Maya Paczuski (1995) had shown in the sand-pile game they had devised that critical states are neither aberrations nor exceptional but rather are natural and inevitable. While avalanches and other phenomena such as volcano eruptions, earthquakes, and bushfires happen—that much can be predicted—just when and where they will occur and their magnitude is, as yet, beyond our ken. Statistically, they do not plot on the normative linearity of a bell curve, but rather show as what is called power law patterns. Every grain of sand that fell in the sand-pile game was the same as the first and each and every ensuing grain but not so its effect. Each gain changed the ontological information of the whole pile which had been changed by each and every grain before it—its effect was in itself *and* in the pile and its history. The potential for avalanches, they found, is "written" into the matter and formation—the information of a pile of sand.

And then there is John Pickering of Warwick University in the UK who, in his 2016 paper, "Is Nature Habit Forming?" asks, "What kinds of things can acquire habits?" He quotes from Charles Peirce's "Guess at the Riddle":

All things have a tendency to take habits. For atoms and their parts, molecules and groups of molecules, and in short every conceivable real object, there is a greater probability of acting as on a former like occasion than otherwise. This tendency itself constitutes a regularity, and is continually on the increase. In looking back into the past we are looking toward periods when it was a less and less decided tendency. But its own essential nature is to grow. It is a generalizing tendency; it causes actions in the future to follow some generalization of past actions; and this tendency is itself something capable of similar generalizations; and thus, it is self-generative. (EP1:277, 1887)

Now this is the kind of thing that gives your average philosopher and physicist the willies and is often why Peirce is rejected by universities here, on the grounds of being a "God-botherer." Pickering (2016) writes:

Even for the open-minded, the proposition that all levels of nature have something mind-like about them is virtually impossible to take seriously at first encounter. In discussions with skeptics, who are the norm, anyone defending panpsychism is likely to be told that "giving atoms minds" or something like that is absurd and un-parsimonious. This is understandable. The only minds human being know are their own (albeit partially) and so the idea that every part of nature is mind-like in some way, panpsychism is easily taken to mean just that—things as simple as atoms are able to make decisions, have thoughts, feelings and so on. (Pickering, 2016: 14).

As we come to understand the process as information making, we come to realize that, quite literally, history matters as a process.

Chapter 5

In the Beginning

To talk about initiation of the universe is to enter into consideration of cosmogony: theories or accounts of the origin (L17) and creation (M18) of the universe. Many shy away from cosmogony because it inevitably raises the issue of God. Lest one blenches at the word "God" may I quote Juliet to Romeo in William Shakespeare's *Romeo and Juliet*:

What's in a name? That which we call a rose
By any other name would smell as sweet. (Shakespeare, II, ii, 1–2)

Nevertheless, it is a necessary precursor to any consideration of cosmology: theories or postulate accounts of the evolution and structure of the universe (M17) and the branch of philosophy or metaphysics that deals with the universe as a whole (M18). Conflating these theories just obfuscates hard but important questions.

In the Western tradition, in response to the scholastics' question *An sit Deus?* (Is there God?), there are three main arguments to the existence of God: the Ontological, the Cosmological, and the Teleological. There are further four classic arguments not raised here: Pascal's Wager, the Moral Argument (deployed by Kant), the Argument from Religious Experience, and the Argument from Miracles, which I have not included here. These are probably all best-called argumentations as they vary depending on who is doing the arguing. For Peirce:

An "Argument" is any process of thought reasonably tending to produce a definite belief. An "Argumentation" is an Argument proceeding upon definitely formulated premisses. (Peirce, 1908)

The Ontological Argument argues from premises that derive from reason alone—from *a priori* and necessary premises—to the conclusion that God exists. It was first proposed by the scholastic Anselm [1034–1109] who in his *Proslogion* derived the existence of God from the concept of a *being than which no greater can be conceived*. Descartes [1596–1650] further developed this in "Meditation V" of his *Meditations on First Philosophy* and this was refined by Leibniz [1646–1716] in *New Essay Concerning Human Understanding*, (1709). The Islamic *falāsifa* (philosopher) Mullā Sadrā's [1571–1640] revision of Avicenna's [980–1037] Cosmological Argument brought a major transition from essentialism to existentialism in Islamic philosophy (*The Metaphysics of Mullā Sadrā. Kitab al-masha'ir (The Book of Metaphysical Penetrations)*) (1992). While he presaged Martin Heidegger's [1889–1976] *Being and Time* (1927) by three centuries, translation of Sadrā's work into English only began in the second half of the twentieth century. Paul Tillich's [1886–1965] Ontological Argument employed central concepts from existentialism and identified God with the ground of all being. Gödel's Ontological Proof, first published by Hao Wang in his *Reflections on Kurt Gödel* (1990) is a logical investigation showing that Leibniz's proof with classical assumptions correspondingly axiomatized is possible.

Peirce would not argue a priori—as discussed, he was vehemently anti-foundationalist—but within mathematics he demonstrated the necessity of his categories of First, Second, and Third as intermediate categories between Aristotle's substance (suchness) and being (thusness) to "realize" the subject and predicate of any hypothesis. From an a posteriori position, beginning with his phenomenology, his arguments, then, were not for *existence* but for *reality*. As he said:

> I myself always use exist in its strict philosophical sense of "react with the other like things in the environment." Of course, in that sense, it would be fetichism to say that God "exists." (PWP.375, 1906)

The Cosmological Argument argues that there was a "first cause," or "prime mover" which is identified as God, while the Teleological Argument argues from order and complexity of the universe and focusing on the plan, purpose, intention, or design, that these are best explained by reference to a creator God. Both the Cosmological Argument and the Teleological Argument are a posteriori arguments and are related. From the perspective of Meno Hulswit's distinction between "*causation*—the production of an effect by its cause" and "*causality*, which is the relationship between cause and effect" (2002: xiv), the Cosmological Argument is concerned with causation and the Teleological Argument with causality.

The Teleological Argument traces back to Socrates [469–399 BCE] who, according to Xenophon [427–355 BCE], argued that the adaptation of human parts to one another, such as the eyelids protecting the eyeballs could not have been due to chance and was a sign of wise planning in the universe. Plato [427–347 BCE] believed that "everything that becomes or changes must do so owing to some cause; for nothing can come to be without a cause" (*Timaeus* 28a). He did not propose creation *ex nihilo* but rather posited a "demiurge" of supreme wisdom and intelligence that made order from the chaos of the cosmos, imitating the eternal Forms. Aristotle [384–322 BCE] introduced his theory of the four *aitia*—prerequisite conditions, rather than causes, for answering "why" questions—which he called the material (*hyle*), formal (*eidos*), efficient (moving), and final (*telos*) *aitia*. For the Stoics, as argued by Chrysippus [280–207 BCE] and commented on by Cicero [106–43 BCE], five of their most basic theses related to causation: (1) the world is ruled by fate, (2) nothing happens without a cause, (3) causation involves exceptionless regularity, (4) causation involves necessity, and (5) there is a fundamental distinction between external and internal causes (Hulswit, 2002: 5–6). Plotinus [205–270], a Neoplatonist, taught that the "One transcendent absolute" caused the universe to exist simply as a consequence of its existence—*creatio ex deo*. His disciple Proclus [412–485] stated that "the One is God." Derived from Proclus's *Elements of Theology* was the *Liber de Causis* (or *Liber Aristotelis de Expositione Bonitatis Purae;* Book of Causes) a Latin translation of an Arabic work whose authorship is unknown, but is considered to be by an Islamic or a Jewish scholar.

In Islamic thought, there are two classical Cosmological Arguments. The first, the *kalām* (the use of reason by believers to justify the basic metaphysical presuppositions of the faith) Cosmological Argument, was developed by Muslim theologians *(mutakallimūm)* such as al-Kindī [801–873] and al-Ghazālī [1058–1111] and is based on the temporality (*huduth*) of the universe. The second, developed by the philosophers (*falāsifa*) such as al-Fārābi (870–950), Avicenna (ibn Sīnā') [980–1037], and Averroes (ibn Rushd) [1126–1198], is based on contingency (*imkan*) of the universe and is atemporal. In addition, there are two teleological arguments—the Qur'anic arguments from design—both of which were developed by Averroes as the argument from providence (*inayah*) and the argument from creation (*ikhtira*). The *mutakallimūm* were influenced by Democritus (b. 480 BCE) who maintained that the universe is comprised of individually indestructible atoms, which account for the formation and dissolution of all observable things. They used reason and argumentation to support their revealed Islamic beliefs and developed the *kalām* version of the argument from the impossibility of an infinite regress. For example, al-Ghazālī argued that everything that begins to exist requires a cause of its beginning. This version of the argument entered the Christian tradition through Bonaventure [1221–74] in his *Sentences* (II Sent. D.1, p.1, a.1, q.2). The *falāsifa* al-Fārābi divided *being* into *necessary* and *contingent*. *Necessary being* is that which exists in itself or that which cannot but exist; nonexistence of it is unthinkable. *Contingent being* is that which receives its existence from another and the nonexistence of which is thinkable or possible. The idea that essence (*Mahiat*) precedes existence (*Wujud*) is a concept that dates back to Avicenna. As mentioned earlier, Avicenna's argument was later revised by Mullā Sadrā. The opposite idea that existence precedes essence was first developed by Averroes.

The Jewish philosopher, Avendehut (ibn Daud) [1110–1180], was the first Jewish Aristotelian. He grounded Jewish theology on the metaphysics of Avicenna, providing an important influence on the work of Maimonides. Maimonides [1135–1204] proposed an elaborate form of Aristotle's unmoved or "prime" mover argument, based on an Aristotelian conception of the motion of the heavens and arguments from the existence of change and from contingency, among others.

Augustine [354–430], originally a Neoplatonist, was a Christian apologist. Like the Jewish scholars, he did not put forward an argument for the existence of God, taking God to be a given, but in Books XI–XIII of *Confessions* [397–8] and Book XI of *The City of God* [413–426] his atemporal teleology out of which the temporal cosmology proceeds by the will (action) of God is both cogent and unique. Hulswit points out that

> most thirteenth century philosophers distinguished two quite different sorts of efficient cause: the *causa prima* and the *causa secunda*, the former being the originative source of being (God), while the latter, which is to be found in created things only, referred to the origin of the beginning of motion or change. Thus, while the (secondary) efficient cause is the source of motion, there is also an active cause that is the source of being. The First Cause works in all secondary causes, which may be considered instrumental causes subservient to the first. (Hulswit, 2002: 2)

Such was the case for the scholastic Thomas Aquinas [1225–1274]. His five arguments for the plausibility of the existence of God begin with our experience of sense objects. Putting aside for the moment, the fourth argument—concerned with qualities—the first three ways are cosmological arguments and are concerned with efficient causes. The fifth way is teleological, dealing, as it does, with the final cause. Hulswit explains:

> The efficient causes are subordinate to the final causes inasmuch as they are a *means* to ends. The final cause is responsible for a twofold necessity in things. The first kind of necessity (I) appears if we consider the cause-effect relationship from the perspective of the efficient cause.... The second kind of necessity (II) appears if the cause-effect relationship is viewed from the perspective of

the effect or end state. It concerns the *means* as necessary condition for the end state. This is called *conditional necessity*. (Hulswit, 2002: 12 & 13)

The Fourth Way by Thomas Aquinas [1225–1274] is commonly called the argument from degree. The argument translated from *Summa Theologicae* into English reads:

> The fourth way is taken from the gradation to be found in things. Among beings there are some more and some less good, true, noble and the like. But "more" and "less" are predicated of different things, according as they resemble in their different ways something which is the maximum, as a thing is said to be hotter according as it more nearly resembles that which is hottest; so that there is something which is truest, something best, something noblest and, consequently, something which is uttermost being; for those things that are greatest in truth are greatest in being, as it is written in Metaph. ii. Now the maximum in any genus is the cause of all in that genus; as fire, which is the maximum heat, is the cause of all hot things. Therefore there must also be something which is to all beings the cause of their being, goodness, and every other perfection; and this we call God. (Aquinas, from *Summa Theologicae*)

The fourteenth century opened with the fullest development of realism in John Duns Scotus (Scottish) [1266–1308] and closed with its eclipse by the nominalism of William of Ockham (English) [1285–1349]. The Renaissance saw a rejection of the Latin of the Scholastics in favor of Greek and, at the hands of philosophers such as Montaigne (1533–1592), a reconsideration of skepticism rather than rational argument. Furthermore, acceptance of nominalism ostensibly ended consideration of formal and final causation which were obfuscated by being sublimated into God's will.

As a prelude to a discussion of causation in modern philosophy, Hulswit summarizes the change in the development of the concept of cause with the rise of modern science in the seventeenth century:

> Explanations by formal causation and final causation being rejected, efficient causation alone was considered to provide rational explanation of the phenomena. Moreover, the concept of efficient causation itself had radically changed. Whereas in the Aristotelian and scholastic tradition (a) efficient causation was not restricted to locomotion, (b) did not involve determinism and (c) efficient causes were conceived as the active initiators of a change, in the seventeenth century the idea took hold that (a) all causation refers exclusively to locomotion, (b) that causation involves determinism and (c) that efficient causes are merely inactive nodes in the chain of events, rather than active originators of a change. . . . Thus, contrary to what is tacitly assumed, the idea that causation involves *determinism* does not have a scientific, but a theological origin. (Hulswit, 2002: 15–16)

Hulswit (2002: 17–45) details the rationalist conceptions of causation of Descartes [1596–1650], Hobbes [1588–1679], Leibniz [1646–1716], Spinoza [1632–1677], and Kant [1724–1804]) and the empiricist approaches of Locke [1632–1704], Newton [1643–1727], Hume [1711–1776], and Mill [1806–1873]. (See Cottingham's *Western Philosophy: An Anthology.*) These approaches support mechanical determinism which purports to explain natural law.

But from whence natural law? In "Taking Science on Faith" (November 24, 2007), the physicist Paul Davies relates his experience:

> When I was a student, the laws of physics were regarded as completely off limits. The job of the scientist . . . is to discover the laws . . . not inquire into their provenance. The laws were treated as "given"—imprinted on the universe like a maker's mark at the moment of cosmic birth—and fixed forevermore. Therefore, to be a scientist, you had to have faith that the universe is governed by dependable, immutable, absolute, universal, mathematical laws of an unspecified origin. . . . If one traces these reasons all the way down to the bedrock of reality . . . only to find that reason then deserts us, it makes a mockery of science. (Davies, November 24, 2007)

This is fideism in a cipher and is unintelligible. Peirce's reintroduction of final causation, recognition of chance, and revision of efficient cause, coupled with acceptance of fallibilism in place of certainty, provide an alternative to blind faith on the one hand and denunciation of God on the other when considering cosmogony and cosmology.

AN SIT DEUS?

To the question *An sit Deus?* (Is there God?) Peirce's response was in the affirmative, but as he stated at the outset in "Answers to Questions Concerning My Belief in God" (CP6.494–521, 1906), "'God' is a vernacular word and, like all such words, but more than almost any, is vague" (PWP.275, 1906).

Nevertheless, although he kept his descriptions of God to a minimum in his metaphysics, thereby reducing the pitfalls of particularization, he supported a form of anthropomorphism. In response to what he, as a pragmaticist, thought of Humanism, Peirce replied: "Why if you had said Anthropomorphism, I should have replied that I heartily embrace most of the clauses of that doctrine" (CP5.536, c.1905). A couple of years earlier he wrote:

> I have after long years of the severest examination become fully satisfied that, other things being equal, an anthropomorphic conception, whether it makes the

best nucleus for a scientific working hypothesis or not, is far more likely to be approximately true than one that is not anthropomorphic. (PAP. 157-8, 1903)

Anthropomorphism is an ascription of human form, attributes, or personality to God. Judaism, Christianity, and Islam are theomorphic religions, that is, these religions hold that, as given in revelation, "God created man in his own image" (Genesis 1:27). Those theomorphists that hold that God is a literal physical being and that mankind is literally created in his image may be termed "anthropotheists." Anthropomorphism should, in the case of Peirce, be differentiated from the concretization of God (gods) of anthropotheism. He was, himself, a Christian for whom God is personal and, in his metaphysics, the capitalized He, Him, and His referred to the categories of First, Second, and Third of the "personality" (see CP6.157, 1891) of the vernacular God. For Peirce, within his semeiotic which informs his metaphysics, "man is a sign" (CP5.314, 1868), that is, "man," too, is vernacular. Probably, though, he was thinking more of man's reason, or mind.

The doctrine of a light of reason seems to be enwrapped in the old Babylonian philosophy of the first chapter of Genesis, where the Godhead says, "Let us make man in our image, after our likeness." It may, no doubt, justly be said that this is only an explanation to account for the resemblances of the images of the gods to men, a difficulty which the Second Commandment meets in another way. But does not this remark simply carry the doctrine back to the days when the gods were first made in man's image? To believe in a god at all, is not that to believe that man's reason is allied to the originating principle of the universe? (CP2.24, c.1902)

And yet, the inescapable gender issues and the nominalism associated with pronouns leave me feeling uneasy with anthropomorphism. The Australian poet Judith Wright expresses this well in her poem *Eve to Her Daughters.*

It was not I who began it.
Turned out into draughty caves,
hungry so often, having to work for our bread,
hearing the children whining,
I was nevertheless not unhappy.
Where Adam went I was fairly contented to go.
I adapted myself to the punishment: it was my life.

But Adam, you know!
He kept brooding over the insult,
over the trick They had played on us, over the scolding.
He had discovered a flaw in himself

and he had to make up for it.

Outside Eden the earth was imperfect,
the seasons changed, the game was fleet-footed,
he had to work for our living and he didn't like it.
He even complained of my cooking
(it was hard to compete with Heaven).

So he set to work.
The earth must be made a new Eden
with central heating, domesticated animals,
mechanical harvesters, combustion engines,
escalators, refrigerators,
and modern means of communication
and multiplied opportunities for safe investment
and higher education for Abel and Cain
and the rest of the family.
You see how his pride had been hurt.

In the process he had to unravel everything,
because he believed that mechanism
was the whole secret—he was always mechanical-minded.
He got to the very inside of the whole machine
exclaiming as he went, So that is how it works!
And now that I know how it works, why, I must have invented it.
As for God and the Other, they cannot be demonstrated,
and what cannot be demonstrated
doesn't exist.
You see, he had always been jealous.

Yes, he got to the centre
where nothing at all can be demonstrated.
And clearly He doesn't exist; but he refuses
to accept the conclusion.
You see, he was always an egotist.
It was warmer than this in the cave;
there was none of this fall-out.
I would suggest, for the sake of the children,
that it's time you took over.

But you are my daughters, you inherit my own faults of character;
you are submissive, following Adam
even beyond existence.
Faults of character have their own logic
and it always works out.

I observed this with Abel and Cain.

Perhaps the whole elaborate fable
right from the beginning
is meant to demonstrate this; perhaps it's the whole secret.
Perhaps nothing exists but our faults?
At least they can be demonstrated.

But it's useless to make
such a suggestion to Adam.
He has turned himself into God,
who is faultless and doesn't exist.
(Judith Wright)

Augustine had said in *The City of God* [413–426], "I have thought that each one, in keeping with his powers of understanding, should choose the interpretation that he can grasp" and yet, anthropomorphizing God has not only *not* increased understanding but has often been positively divisive and detrimental regarding cosmogony and cosmology and especially so since the ascendency of nominalism.

While as an agnostic one cannot *know* either a priori, that is, from introspection or axioms, or a posteriori, from revelation or deduction, in accord with Peirce's pragmaticist metaphysics, the *reality* of God is meaningful: insofar as God is general, the principle of excluded middle does not apply, nor, in being vague, does the principle of contradiction apply (see CP 5.505, c.1905). God is not constrained by existence; as *real*, God is boundless.

QUID SIT DEUS?

In response to the scholastics' question *Quid sit Deus?* (What is God?), Peirce spoke of creative activity. He spoke of God in general and not God in particular—of the reality of God—of God as universal. God is a sign.

Hausman brings to our attention that

In 1903, Peirce compared his sense of the word "category" with his understanding of the senses of the word for Aristotle, Kant and Hegel. He distinguished two orders of categories. . . . The first order of categories is particular. Particular categories are restricted in their range of application; they form a series, only one member of which is present in or dominant in any one phenomenon. The second order is universal and these universal categories, to which Peirce limited himself, are all present in all phenomena. (Hausman, 1979: 203–4)

During the 1880s he spent time developing his evolutionary cosmology and in a draft *Sketch of New Philosophy* (MS.928, n.d.) he described three kinds of philosophy: the elliptic, parabolic, and hyperbolic philosophies, a nomenclature based on Felix Klein's (1849–1925) triple distinctions of geometries ("elliptic" for Riemann's [1826–1866] geometry on a surface of positive curvature; "parabolic" for Euclidean [365–275 BCE] geometry; and "hyperbolic" for Lobachevsky's [1792–1856] geometry). In a letter to Christine Ladd-Franklin (CP8.317, 1891), describing his cosmology that he had developed over the previous ten years, he wrote of his theory that "the evolution of the world is hyperbolic, that is, proceeds from one state of things in the infinite past, to a different state of things in the infinite future."

Space, I take to be the nothing that is room to move. Peirce argued (CP6.212, 1898) that "the whole nature and function of space refers to Secondness. It is the theatre of the reactions of particles and reaction is Secondness in its purity." From nothingness comes possibility; from possibility comes potentiality which is chaos. In accord with this, as Peirce argued:

> The very first and most fundamental element that we have to assume is a Freedom, or Chance, or Spontaneity, by virtue of which the general vague nothing-in-particularness that preceded the chaos took a thousand definite qualities. The second element we have to assume is that there could be accidental reactions between those qualities. The qualities themselves are mere eternal possibilities. But these reactions we must think of as events. Not that Time was. But still, they had all the here-and-nowness of events. (RLT.260, 1898)

The law of chance is lawlessness—it is random—or as Anderson (1987: 100) puts it, "chance's reality is the sheer indeterminacy displayed by the original chaos in relation to a future universe." Because present-day physicists are now able to identify momentary secondness as particles—photons for the electromagnetic field, gravitons for the gravitational field, and so on—that pop into and out of existence from nothingness, degenerate thirdness is identifiable. That modern physicists have discovered that the vacuum—what was in Peirce's day known as the ether—is a seething ferment of activity makes chance no more determinate than it was prior to the discovery. Anderson (pp. 102–3) notes: "When we view chance from the side of actuality rather than possibility, we call it spontaneity." Of itself, chance is not necessary, "chance may be a necessary condition of creative evolution, but it is not a causal source" (p. 102). In saying that "Tychism, or the doctrine that absolute chance is a factor of the universe" (RLT.260, 1898) Peirce was not defending Tychasticism, the doctrine that indeterminism is the only factor in the universe. Out of chance comes a tendency toward regularity.

Despite the excitement with which Peirce greeted the publication of Charles Darwin's *Origin of Species* (1859) he found the theory to be mechanical. Though Darwin had flirted with William Paley's [1743–1805] analogy between the operation of nature and the movements of a well-designed watch, often regarded as a classic statement of the Teleological Argument for God's existence, he ultimately opted for a *law* of natural selection. The underlying belief of both Paley's analogy from design and Darwin's theory is predetermined—by God for Paley and law for Darwin—and therefore fatalistic.

What sets Peirce's cosmogony and cosmology apart from others in Western thinking, however, is his idea of developmental telos. In contradistinction to the Teleological Argument of Augustine (and of the *falāsifa* and Aquinas who followed his lead), in which the plans and design, including the laws of the universe, were set in place by God before creation, Peirce's teleology was evolutionary. In this it was and is truly radical and revolutionary. Referring back to his defense of anthropomorphism, despite my misgivings, his description of personality as "some kind of coordination or connection of ideas" does serve to clarify his evolutionary telos.

> This personality, like any general idea, is not a thing to be apprehended in an instant. It has to be lived in time; nor can any finite time embrace it in all its fullness.
>
> But the word coordination implies somewhat more than this; it implies a teleological harmony in ideas and in the case of personality this teleology is more than a mere purposive pursuit of a predeterminate end; it is a developmental teleology. This is personal character. A general idea, living and conscious now, it is already determinative of acts in the future to an extent to which it is not now conscious.
>
> This reference to the future is an essential element of personality. Were the ends of a person already explicit, there would be no room for development, for growth, for life; and consequently there would be no personality. The mere carrying out of predetermined purposes is mechanical. (CLL233–4, 1891)

As a creative activity, God's *finious* cause is "becoming" the *summum bonum*; it is growth itself. This, Peirce named *agapastic evolution* or *agapasm*. Evolution by fortuitous variation, he called *tychastic evolution* or *tychasm*; evolution by mechanical necessity he labeled *anacastic evolution* or *anancasm*.

> All three modes of evolution are composed of the same general elements. Agapasm exhibits them the most clearly. The good result is here brought to pass, first, by the bestowal of spontaneous energy by the parent upon the offspring and, second, by the disposition of the latter to catch the general idea of those about it and thus to subserve the general purpose. In order to express the relation that tychasm and anancasm bear to agapasm let me borrow a word from geometry.

An ellipse crossed by a straight line is a sort of cubic curve; for a cubic is a curve which is cut thrice by a straight line; now a straight line might cut the ellipse twice and its associated straight line a third time. Still the ellipse with the straight line across it would not have the characteristics of a cubic. It would have, for instance, no contrary flexure, which no true cubic wants; and it would have two nodes, which no true cubic has. The geometers say that it is a degenerate cubic. Just so, tychasm and anancasm are degenerate forms of agapasm. (Peirce, 1893)

Although, as Anderson puts it, "the meaning of God's telos is precided by the specific actualization which fix it as its own referent" (1987: 113), the specific actualization, although irreversible, is not static but continues into the future. As created, the universe is being and to that extent it is fact; as referent, it has been created tychasticly, anacasticly, and agapasticly and continues in like manner creating. It is development, as it is commonly understood, as continuity. In general, this may be considered as recreation, it is a change and in changing changes and is changed. In his paper "Design and Chance" Peirce explained:

Suffice it to say that as everything is subject to change everything will change after a time by chance and among those changeable circumstances will be the effects of changes on the probability of further changes. And from this it follows that chance must act to move things in the long run from a state of homogeneity to a state of heterogeneity. (W4:550, 1883–4)

Because when *analyzing* the strands of Peirce's realism one can lose sight of the fact that not only are his three categories of First, Second, and Third irreducible to each other, so too is realism irreducible to monad or dyad; only as triad is it wholly intelligible. Here is the philosophical underpinning of modern theories of quantum, emergence, and complexity, the antithesis of the principles of universal determinism, reductionism, and disjunction, the hallmarks of much of twentieth-century philosophy and science. Yet when it came to understanding these concepts, Peirce, along with other philosophers of the late nineteenth and early twentieth century who propounded alternatives to mechanistic science, including Samuel Alexander [1859–1938], Henri Bergson [1859–1941], and Alfred North Whitehead [1861–947], were rejected as resorting to *deus-ex-machina* (god from the machine) explanations.

In their introduction to volume 4 of the *Exploring Complexity* Series (2008) Alicia Juarrero and Carl Rubino bring to our attention (p. 4) that

the term *emergence,* first proposed in the 1870s by George Henry Lewes (1817–1878) in *Problems of Life and Mind* and then taken up by Wilhelm Wundt (1832–1920) in his *Introduction to Psychology,* was coined precisely to identify instances in chemistry and physiology where new and unpredictable properties appear as products that are emphatically not the mere sum of the separate elements from which they arise.

Yet, as Juarrero and Rubino point out:

> Not until Ilya Prigogine was awarded the Nobel Prize in 1977 for his work on Dissipative Structures and, with Isabelle Stengers, published the surprisingly popular *La Nouvelle Alliance* two years later, did many serious scientists and philosophers dare question "the goals, methods and epistemology" of modern science. Doing so required scientists to reconsider the creative aspects of nature, made manifest in an evolutionary process displaying irreducibly emergent properties. (Juarrero & Rubino, 2008: 7)

Prigogine had read Peirce and credited him with anticipating the "new" physics. Notwithstanding this and other high-profile endorsements of him, Peirce is still largely sidelined. Although the "new physics" is progressing, development of the philosophy that informs it is sadly lagging. The astrophysicist Erich Jantsch, inspired by and drawing on the work of Prigogine concerning dissipative structures and nonequilibrium states, explored in his book *The Self-Organizing Universe: Scientific and Human Implications of the Emerging Paradigm of Evolution* (1980), the idea of self-organization as a unifying evolutionary paradigm. Nevertheless, Jantsch's theory argued that God is the self-organizing dynamic of the cosmos and in this it was considered irrational, as was the theory of Fritjof Capra, a member of the Fundamental Fysiks Group at Berkeley University, with its talk of cosmic mind and self-organization.

Such rejection is prejudicious. It brings to mind a quotation made famous following the 1939 publication of the first edition of the book *Alcoholics Anonymous* and attributed to Herbert Spencer:

> There is a principle which is a bar against all information, which is proof against all arguments and which cannot fail to keep a man in everlasting ignorance— that principle is contempt prior to investigation. (Alcoholics Anonymous, 1939)

Although differently placed in succeeding editions of the book, the quotation has always appeared in the more than 20 million copies printed since its original publication. In a paper "The Survival of a Fitting Quotation" Michael StGeorge (2002) revealed: "Herbert Spencer never wrote or said anything resembling this quotation." It was first published as quoted earlier by a Canadian named Rev. William H. Poole who was arguing that the Anglo-Saxon race is actually descended from the ten lost tribes of Israel in his book *Anglo-Israel or, The British Nation: The Lost Tribes of Israel* (1879). Poole attributed the quotation to William Paley. What Paley in fact wrote in *A View of the Evidences of Christianity* (1794):

The infidelity of the Gentile world and that more especially of men of rank and learning in it, is resolved into a principle which, in my judgment, will account for the inefficacy of any argument, or any evidence whatever, *viz.* contempt prior to examination. (Paley, 1794)

StGeorge says: "In this context, Paley was trying to give reasons why the Christian faith was rejected by the ancient Greeks and Romans." This is the same William Paley mentioned earlier in relation to the Teleological Argument to the existence of God, which Darwin rejected but which would be attractive to contemporary adherents of Intelligent Design. In his *Natural Theology* (1802) Paley presented a watchmaker analogy to support his argument. His analogy is strikingly similar to the sundial/water-clock one reported by the Roman statesman Cicero (106–43 BCE) as presented to him by the Stoic philosopher, Quintus Lucilius Balbus (fl. 100 BC) (*De Natura Deorum Liber Secundus*) but perhaps not surprisingly, Paley made no acknowledgment of any source.

Dismissing Alexander, Bergson, and Whitehead as resorting to *deus-ex-machina* explanations misrepresents them. The phrase *deus-ex-machina* is a *plot device* whereby a seemingly unsolvable problem is suddenly and abruptly solved with the contrived and unexpected intervention of some new event, character, ability, or object. A deeper investigation into the work of these three, however, would show them to be panentheistic.

The word panentheism (from Greek meaning "all-in-God") was coined by Karl Christian Friedrich Krause in 1828. Panentheists see God and the world as interrelated in process, with the world being in God and God being in the world. Panentheism differentiates itself from pantheism, which holds that everything composes an all-encompassing, immanent God or that the universe (or Nature) is identical with divinity. Pantheists do not believe in a personal or anthropomorphic god. Based on the work of Baruch Spinoza, whose treatise, *Ethics*, was an answer to Descartes's dualist theory that the body and spirit are separate, pantheists hold the two are the same.

Michael Raposa, in *Peirce's Philosophy of Religion* (1989) suggests that Peirce was a panentheist.

Peirce, while definitely not a pantheist, might be properly labeled a panentheist, that is, one who views the world as being included in but not exhaustive of the divine reality. Such a view neither undermines the doctrine of creation nor collapses the distinction between God and the universe. (Raposa, 1989: 51)

In a note to this passage Raposa comments:

The term "panentheism" is used here only in the broadest possible sense and not to designate a specific doctrine associated with a particular thinker or group of thinkers. Decades ago Charles Hartshorne noted Peirce's panentheistic tendencies regretting only that he "falls short" of embracing the dipolar God of contemporary process philosophy, clinging instead to a more classical theism...But in arguing that these classical elements are inconsistent in Peirce's own system, Hartshorne does not appear to have assessed carefully enough what Peirce had to say about continua and their singularities, about the logic of vagueness, or about the *semeiotic* relationship between God and the universe. (Raposa, note 22: 160-161)

Notwithstanding Raposa's qualification, in the spirit of Peirce naming his version of pragmatism "pragmaticism" and Mayorga (2007) calling his realism "realicism," let me coin a term to account for Peirce's hypothesis of God and call it "pantheisticism." As with "pragmaticism" and "realicism," "pantheisticism" is a term of differentiation—in this case differentiated from pantheism and panentheism. The most obvious difference is that while "pantheism" is monadic and "panentheism" is dyadic, "pantheisticism," being attributable to the Peircean idiom, is triadic. In a very early manuscript of fragments from a *Treatise on Metaphysics* (MS.921, 1861) Peirce wrote of idealism, materialism, and what he called realistic pantheism as representing the three worlds of mind, matter, and God—worlds which both mutually exclude and include each other. I found no evidence that he employed this term again. Neither, even having read the German philosopher Krause [1781–1832], did he use the term panentheism, despite Krause describing it as universal and idealistic. Instead, inspired by Schilling [1775–1854] whose philosophy, like his, "avoids all and every sort of dogmatism" (Ibri, 2009: 282) he advanced "Objective Idealism" in the first of his *Monist* series of 1891–1893 on metaphysics, "The Architecture of Theories" (CP6.7–34, 1891). "But," as he wrote,

before this can be accepted it must show itself capable of explaining the tri-dimensionality of space, the laws of motion and the general characteristics of the universe, with mathematical clearness and precision; for no less should be demanded of every philosophy. (CP6.25, 1891)

Back to "the beginning": as already stated, Peirce was a Christian. Raised a Unitarian—so named because of its disavowing the Trinity doctrine—he converted to the Episcopalian church as a precondition of his first marriage in 1863. Some relevant notable Unitarians were John Locke [1632–1704] who is also claimed by the Anglicans (Episcopalians), Ralph Waldo Emerson [1803–1882], Peirce's peer and friend Francis Ellingwood Abbot [1836–1903], John Dewey [1859–1952] best known of the classical pragmatists, Arthur Lovejoy [1873–1962] negative commentator of Peirce's philosophy, Alfred North Whitehead [1861–1947] who is listed as a Unitarian friend, and Charles Hartshorne [1897–2000] who, despite having years of access to his papers, never did "get" Peirce.

Lest Peirce's trichotomy be attributed to his conversion, it should be noted that in 1913 in an unpublished document (MS.681) he wrote that he did not know and had never inquired whether there was any connection between his own trichotomy and the Divine Trinity. As previously stated, he was scathing about most theologians and dismissive of parts of the Bible, but he did show respect for and paid attention to John who, I suggest, was Peirce's muse for his cosmogony. The Gospel according to John opens (John 1: 1-3) in the King James edition of the Bible:

> In the beginning was the Word and the Word was with God and the Word was God. The same was in the beginning with God. All things were made by him; and without him was not anything made that was made. (John 1: 1-3, King James edition of the Bible)

I prefer the New English Bible translation (1970) because it uses "created"—which resonates with originality—rather than "made"—with its overtones of "manufactured."

> When all things began, the Word already was. The Word dwelt with God and what God was, the Word was. The Word, then, was with God at the beginning and though him all things came to be; no single thing was created without him. (John 1: 1-3, New English edition of the Bible)

Now I find this a remarkable statement which, together with another remarkable statement by John that "God is love" (4: 8), is a profound cosmogony.

This is just one of hundreds of creation myths. A myth is a *form* for expressing profound truths and not a fiction, as detractors would have when denigrating others and promoting their own version. If one considers basic types of creation myths, then Peirce's is a combination of three types: *ex nihilo* (out of nothing), creation from chaos, and emergence. Being both an expression of truth and yet speculative, within Peirce's architectonic myth is metaphysical. Unlike Topsey's version of her own beginning, the universe did not "just grew"; from Peirce's cosmogony—what I have called his "pantheisticism"—flows his cosmology which details the trichotomy of an evolving and developing telos.

Floyd Merrill, in his paper "Overdeterminacy, Underdeterminacy, Indeterminacy," (2005) explores how Peirce's semeiotic "reflects a tension and potential mediation between vagueness and generality, the individual and the universal and discontinuity and continuity, as well as between self and other and self and sign, in such a manner as to defy precise description." Yet, as he further explains:

Taking into account the composite characteristics of possibility (Firstness), actuality (Secondness) and potentiality (Thirdness), a certain "Principle of Indeterminacy" is crucial to an understanding of Peirce's notion of semiosis. (Merrill, 2005)

From the vantage point of "there now" inquiry can be objective: in the "here now," a sentence can be determinately judged either "true" or "false" but is atemporal and static. When considered from "here then" and "there-then," however, its value, having embraced temporality, will have suffered change, if only in time. In this sense, vagueness and generality are complementary forms of indeterminacy.

From his limited understanding of philosophy, the Nobel Prize–winning theoretical physicist Frank Wilczek in his book explains:

Philosophical realists claim that matter is primary, brains (minds) are made from matter and concepts emerge from brains. Idealists claim that concepts are primary, minds are conceptual machines and conceptual machines create matter. . . . Both can be right at the same time. They describe the same thing using different language. (Wilczek, 2008: 112)

Both Frank Wilczek and Paul Davies have admitted to being unfamiliar with Peirce's work. My reading of and meeting with both these great physicists (neither of whom take issue with God) convinces me that they would gain as much from Peirce as I have from them. The value of Peirce's three irreducible categories and his perennial classification of the sciences—the very weft and the weave of his system—is that in enabling us to make our ideas clear, the reality of "mattering" is realized.

Chapter 6

Cosmology

The emphasis for cosmogony is on Firstness, for cosmology it is on brute force of Secondness. Science has identified four forms of power: electromagnetism, the strong nuclear force, the weak nuclear force, and gravity. The first three of these are accounted for in the Standard Model of Particle Physics. See Figure 6.1 (below).

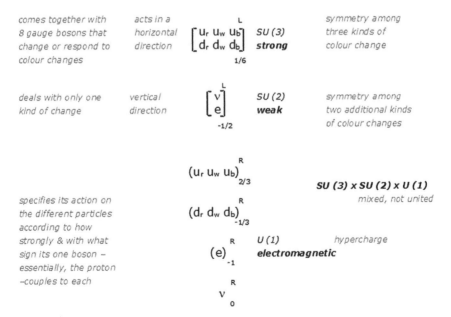

Figure 6.1   The Organization of Particles and Interactions in the Standard Model of Particle Physics. *Source*: Adaptation of Wilczek (2008: 164 and annotated).

83

Frank Wilczek makes the quantum world intelligible and accessible to non-physicists like me. For him (and others of like mind) the *summum bonum* is supersymmetry (SUSY). It "unifies the three fundamental forces of the standard model, suggesting that electromagnetism and the strong and weak nuclear forces merge into a single superforce at high energies." "But what if supersymmetry is wrong?" asks Amanda Gefter (2011). Experimental physicists at CERN's Large Hadron Collider have so far had little success in their search for SUSY, yet for Wilczek, "SUSY is too beautiful to be wrong." Nevertheless, should substantive evidence of an alternative theory emerge rather than for SUSY, Wilczek would be disappointed, but as the consummate scientist, he would concede.

Gravity is not accounted for by the Standard Model. Many believed that the missing link was the particle dubbed the Higgs Bosom which, it was theorized, gives matter its mass. The Higgs was discovered in July 2012 but the study of the data has shown that it doesn't meet expectations. Though I have read a number of reports by naysayers of various stripes that condemn SUSY to the graveyard of defunct theories (see Michael Slezak's 2012 article "Higgs Boson Is Too Saintly and Supersymmetry too Shy") I've yet to see anything published by supersymmetry supporters. Meanwhile, alternative theories such as string theory have so far been unable to design experiments for testing their theories empirically.

Physicists the world over are looking for what is called "the theory of everything" which they believe would unify the quantum description of the three known forces: the electromagnetic and the strong and weak nuclear forces—the Standard Model—with Einstein's theory of gravity. In spite of the hubris of the claim that any theory can explain everything and thereby, according to some such as Stephen Hawking (Hawking & Mlodinow, 2010) can explain away God, many theoretical physicists *do* work cooperatively. Together with experimental physicists they design and conduct experiments to test their theories, they share data and they admit to the hypothetical nature of their findings. In this regard they reflect Peirce's vision of a scientific community. Furthermore, though most are not aware of this, they come closest to following his model for inquiry according to his perennial classification of the sciences. Michael Duff, chair of Theoretical Physics at Imperial College London, explains in his *New Scientist* article (2011) that whether engaged in explaining the data generated by their experimental colleagues or predicting phenomena that have not yet been discovered, the work of theoretical physicists is grounded in mathematics. This is not a mathematics founded on logic. As Claudine Tiercelin (2010: 111) put it, "The real opposition between logic and mathematics lies between the theoretical or observational aspect of inference on the one hand and its practical or operational part on the other." As she explains:

> Peirce's central idea about necessary deductive reasoning is, indeed, that it proceeds by construction of diagrams which are species of icons, the essential

feature of which is to be able to represent the *formal* sides of things, so that they have less a function of resemblance to their objects than of exemplification or exhibition. Also they are formal and not mere empirical images. (Tiercelin, 2010: 100)

For Peirce, "the essence of mathematics lies in it making pure hypotheses and in the character of the hypotheses which it makes" (CP3.558, 1989). In both the case of predicting phenomena that have not yet been found, such as the case of Einstein's general theory of relativity or in explaining what has been discovered by experimental physicists, as with much of quantum theory, what is being observed is the product of the phaneron, that is, what "is present to the mind in any sense or in any way whatsoever, regardless of whether it be fact or figment" (CP8.213, c.1905). Both call for the formation of a hypothesis, that is, abduction. Inference to the best explanation (IBE) is the commonly held understanding of what is meant by abduction and such a meaning may be appropriate where what is presented to theoretical physicists is fact or data requiring explanation. Abduction as pure hypothesis making is what is called for in such cases as Einstein's general theory of relativity, first present to his mind as figment, that is, in his imagination. This requires abduction, not as IBE but as possibility. Peirce, arguing for originality in mathematics said, "Originality is not an attribute of the *matter* of life, present in the whole only in so far as it is present in the smallest parts, but is an affair of *form*, of the way in which parts none of which possess it are joined together" (CP4.661, 1908). According to Michel Duff (2011), though imagination, or speculation as he calls it, is a vital part of the scientific process:

It is always constrained by the straightjacket of mathematical consistency and compatibility with established laws. Even before it was tested experimentally, Einstein's theory of general relativity had to pass several theoretical tests. It had to yield special relativity and Newtonian mechanics in those areas where they were valid, as well as predict new phenomena in those where they were not. (Duff, June 4, 2011)

This is as maybe; however, Professor Reginal Cahill, head of the Process Physics research group once at Flinders University, refuted Einstein's non-process model.

The ill-conceived attempt to formulate a model of reality in which absolute motion was without meaning resulted in the introduction of the space-time construct. This is a four-dimensional geometrical construct in which the one-dimensional model of time is fused with the three-dimensional model of space, but fused in a special way in that observers in relative motion would identify different foliations of the construct as their geometrical time lines, so that their modellings of time no longer coincided and as a consequence they could no longer necessarily agree on the time-ordering of events. Rather than being seen as an indicator of something wrong in the model this aspect of the space-time model became a celebrated feature and the whole notion of change, of the evolution of reality from a past state to a future state disappeared and reality, it was claimed, was simply a frozen unchanging four-dimensional block of geometry: when the universe was formed the whole of the *future* of that universe also popped into existence. So what about our experiences of the *present moment* and the distinction between past and future? Well that was dismissed as being some trick of our minds and a trick that psychologists should investigate, but certainly not physicists. (Cahill, 2003: 10–11)

In contrast to the reality paradigm associated with the names of Galileo, Newton, Einstein, and many others, the Process Physics group was developing an information-theoretic modeling of reality out of which a new theory of gravity emerged that is in agreement with all experiments and observations. According to Cahill (2005: 15), "the self-referentially limited neural network model, that underpins process physics, is remarkably akin to Peirce's effete mind."

Speaking of this model Christopher Klinger, in the abstract of his thesis *Process Physics: Bootstrapping Reality from the Limitations of Logic* (2005), says:

> Process Physics utilizes the limitations of logic first identified by Gödel and asserts the priority of process and relational endophysics, realized via a stochastic, autopoietic bootstrap system whose properties emerge *a posterior* rather than being assumed *a priori*. (Klinger, 2005)

It is worth exploring this statement beginning with the idea of Gödel's Incompleteness Theorems (1931) which revealed the limitations of logic.

> Gödel . . . proved fundamental results about axiomatic systems, showing in any axiomatic mathematical system there are propositions that cannot be proved or disproved within the axioms of the system. In particular the consistency of the axioms cannot be proved. This ended a hundred years of attempts to establish axioms which would put the whole of mathematics on an axiomatic basis. One major attempt had been by Bertrand Russell with *Principia Mathematica* (1910–13). Another was David Hilbert's formalism which was dealt a severe blow by Gödel's results. The theorem did not destroy the fundamental idea of formalism, but it did demonstrate that any system would have to be more comprehensive than that envisaged by Hilbert. Gödel's results were a landmark in 20th-century mathematics, showing that mathematics is not a finished object, as had been believed (Cahill, 2005).

Peirce would agree for it vindicates his rejection of logicism. I believe he would also be open to the practice of endophysics, which is, in accord with pragmatism, the philosophy of experience. Kirsty Kitto in "Quantum Theories as Models of Complexity" explains:

> The term endophysics [Otto Rössler] has been coined to refer to the study of systems which have observers enclosed within them. Such a view is participatory; how we look determines what we see and with the development of new internal perspectives we might begin to understand the way in which the context of such an observer can affect the observations they make of the system itself. (Kitto, 2007: 10)

Endophysics is a methodology that is concerned with determining meaning, in contrast to exophysics with its epistemological focus on finding truth where observers, in the name of objectivity, are outside looking in. The commonly held understanding of what is meant by objectivity, which equates "disinterest" with "disengagement," is plain wrong: reality is real regardless of you or I. Werner Heisenberg once remarked that "the division of the world into an objective and a subjective side much too arbitrary" (1971: 88).

Peirce would not only have no argument with the stochastic approach taken by Process Physics but he championed it himself. Stochastic (from Greek *meaning* "skilled at aiming," since *stochos* is a target) describes an approach to anything that is based on probability. In probability theory, a stochastic process is the counterpart to a deterministic process. Instead of dealing with only one possible reality of how the process might evolve under time, in a stochastic process there is some indeterminacy in its future evolution described by probability distributions. This means that even if the initial condition is known, there are many possibilities the process might go to, but some paths may be more probable and others less so.

Finally, Peirce would have welcomed the theory of autopoiesis, so akin to his evolutionary cosmology of *tychism*, *synechism*, and *agapasm*. An autopoietic (Humberto Maturana 1928–2021) system is one which produces itself. It is a system of components that interact according to some network of interrelationships, where the network of interrelationships is produced by the system components themselves and is the process of dynamics self-maintenance. Autopoietic sets interact in such a way that while the components may be replaced, the system as a whole keeps the same cohesive structure.

A comparison of the models of Newton, Einstein, and Process Physics can be found at Figure 6.2 (below).

The Process Physics model, in rejecting Einstein's static model, accounts for the reality of absolute motion, of a complex evolving universe and of space and time. According to Cahill (2003:15) Einstein's error in rejecting absolute motion trapped twentieth-century physics in the non-process or no

Theory	Time	Space	Gravity	Quantum	k2
Newton	geometry	geometry	force	Quantum Theory	n3
Einstein	curved geometry		curvature	Quantum Field Theory	0
Process	process	quantum foam	inhomogeneous flow	Quantum Homotopic Field Theory	n(n2 i 1)

Figure 6.2 Comparisons of Newtonian, Einsteinian, and Process Physics. *Source*: Adaptation of Cahill (2003:54).

now mindset. He relates Rudolf Carnap (in P. A. Schilpp, Ed.) *The Philosophy of Rudolf Carnap* (1963, 37):

> Once Einstein said that the problem of the Now worried him seriously. He explained that the experience of the Now means something special for man, something essentially different from the past and the future, but that this important difference does not and cannot occur within physics. That this experience cannot be grasped by science seems to him a matter of painful but inevitable resignation. I remarked that all that occurs objectively can be described in science: on the one hand the temporal sequence of events is described in physics; and, on the other hand, the peculiarities of man's experiences with respect to time, including his different attitude toward past, present and future, can be described and (in principle) explained in psychology. But Einstein thought that scientific descriptions cannot possibly satisfy our human needs; that there is something essential about the Now which is just outside of the realm of science. (Cahill, 2003: 17–18)

Notwithstanding my discomfort with calling anything "absolute," Process Physics restores time's arrow and *haecceity* to a complex evolving universe.

Coming back down to earth: Jon Ogborn and Edwin F. Taylor in their article "Quantum Physics Explains Newton's Laws of Motion" (*Physics Education* January 2005: 26–34) explain that Newton was obliged to give his laws of motion as fundamental axioms, but we now know that the quantum world is fundamental and Newton's laws can be seen as *consequences* of fundamental quantum laws.

> Newton's law fixes the path so that changes in phase from changes in kinetic energy exactly match those from changes in potential energy. This is the modern quantum field theory view of forces: that forces change phases of quantum amplitudes. . . . What Newtonian physics treats as cause and effect (force producing acceleration) the quantum "many paths" view treats as a balance of changes in phase produced by changes in kinetic and potential energy. . . .The

structure of our world and our observation of it both depend on this difference between the group behaviour of photons and the group behaviour of electrons. (Ogborn & Taylor, 2005: 33–34)

Thus far I have given theoretical accounts of ordinary cosmological mattering but there are two further types of matter that physicists have discovered. These are dark matter and dark energy. They are referred to as "dark" because they do not absorb or reflect light; they are transparent. Data from the Plank satellite announced in March 2013 gave the composition of matter in the cosmos as 5 percent ordinary matter, 26 percent dark matter, and 69 percent dark energy.

The Swiss astronomer Fritz Zwicky [1898–1974], while examining the Coma galaxy cluster in 1933, was the first to use the virial theorem to infer the existence of unseen matter, which he referred to as *dunkle Materie*, "dark matter." The existence and properties of dark matter are inferred from its gravitational effects on visible matter, radiation, and the large-scale structure of the universe. Zwicky's *dunkle Materie* was ignored for the next thirty years, but over the past fifty years it has been intensely studied both theoretically and empirically. Dark matter has still to be directly identified, but there is a sense among physicists that such an event is imminent. Katherine Freese's *The Cosmic Cocktail* (2014) gives a thorough insider account of dark matter.

What though of dark energy? In the mid-90s cosmologists had had the sense that they'd painted the Universe into a corner: they had a good handle on the numbers that described it and it remained only to dot a few "i"s and cross a few "t"s. That complacency was blown out of the water in 1998, when "dark energy" burst on the scene. Saul Perlmutter (2001) writes about it: https://www.slac.stanford.edu/econf/C990809/docs/perl-mutter.pdf

The Supernova Cosmology Project team, headed by Saul Perlmutter, along with the competing High-z Supernova Search Team led by Adam Riess and Brian P. Schmidt, found evidence of the accelerating expansion of the universe based on observing Type *Ia* supernova in the distant universe. Dark energy was inferred from this discovery along with a further discovery by several teams that there is much more mass in the universe than could be accounted for by normal matter alone.

What we don't know, we have not only to discover but work to understand. Those of us, such as myself, who are not cosmologists, are unable to work directly on such matters; we are reliant on the physicists to convey their findings to us in such a way that we can understand. I do not understand dark energy because the physicists do not yet understand it. What I trust (this far) is that both dark matter and dark energy are necessary in the process of the self-creating universe.

This chapter, Cosmology, explores theories or postulate accounts of the evolution and structure of the cosmos. Along with Peirce, I take freedom—creativity, spontaneity, chance—as first principle.

Existential philosophy also begins from freedom out of nothingness. The French Existentialist Jean-Paul Sartre [1905–1980] in his major philosophical work of ontology, *l'Être et le néant* (*Being and Nothingness*) (1943), offered an account of existence in general, including being-for-itself by which humans engage in independent action. His account included discussion of the human propensity to refuse responsibility for one's own actions as "fear of and flight from freedom" which he condemned as *mauvaise foi* (bad faith). Unfortunately, he did not follow up with a coherent metaphysics.

The Algerian-French writer Albert Camus [1913–1960] struggled throughout his oeuvre with reconciling the ideology of existentialism with existential reality, against a backdrop of the Western tradition of Cartesian bivalence and mechanical determinism. As a *pied noirs* (the name denotes a French citizen born and living in Algeria) he was raised in a deeply imperialist and colonial culture. His struggle between notions of justice and equality forced him, toward the end of his life, to forgo his tenacious ahistorical frame of reference, but he was killed in a car accident in 1960 before reconciling this with his long-held beliefs.

Raymond Boisvert of Siena College, Loudonville, NY, explores this idea in his paper "Camus: Between Yes and No" (2013). While he agrees that Camus had not achieved a resolution before his untimely death, he shows the path and indications of this purpose throughout Camus's writings. Camus introduced two related ideas of particular merit that took his work beyond the ideological position of Sartre's Existentialism. The first of these was the notion of limit or measure. Boisvert notes:

> The radical freedom of the existentialists may have ignored this notion; but limits are, as Heraclitus pronounced, woven into the fabric of things. Woe to those who would transgress those limits. (Boisvert, 2013. 10)

The second idea was what Camus called *consentment*. Contrary to bivalence of "yes" *or* "no," Camus stretched toward wedding these opposites in human affairs into "yes" *and* "no." Boisvert concludes:

> Camus has projected a comprehensive series of work divided into three cycles. The first two, the cycle of the absurd and the cycle of revolt, he got to accomplish. The third could have taken him into a realm where an overarching "yes" nonetheless incorporates the need to struggle associated with a "no." He called it the cycle of love. (Ibid.)

Thus, beginning from freedom and bringing together Peirce's three irreducible categories and his architectonic and Camus's recognition of limit and his idea of consentment, let me sketch an intelligible, informed cosmology.

I cannot begin "once upon a time," unfortunately, much as I'd like to. Before the beginning was nothing—not even time. Nothing is full of potential which too is nothing as Firstness alone. So too is chance, as Firstness, nothing without Secondness and Thirdness. That which is called chaos came from nothingness, by way of spontaneous chance coupled with the powered potential "to be" or "to matter," which is Secondness. Without Thirdness, chaos is, as its name denotes, disorder. But when purpose is seen as inherent to being—as the telos of "mattering"—then "mattering" as Thirdness, which is evolution, begins: out of nothing emerges chaos; from chaos the universe creates itself.

Notwithstanding that the physicists' theory of the "birth" of universe, the Big Bang, was from a point and that a point is by no means nothing, along with the physicists, I cannot (yet) describe in their terms, the origins of that point. Therefore, being mindful that the point was itself "mattering," I will follow from there with my suggestion that the evolution of the laws of physics can be theorized by reference to Peirce's pragmaticism and Camus's ideas of "limit" and "consentment." The object of "mattering" is an event and therefore I suggest that maybe the first limit or measure to evolve was time. The physicists tell us, for example, that one second after the Big Bang the universe was filled with neutrons, proton, electrons, anti-electrons, photons, and neutrinos. During the first three minutes, the light elements—deuterium, an isotope of hydrogen, most of which combined to make helium and trace amounts of lithium—were generated. However, according to France's National Centre of Space Research, for the first 380,000 years or so, the universe was essentially too hot for light to shine. I expect this is the horizon beyond which we are unable to see and that therefore will never be able to *see* the Big Bang.

The patterns which we call natural law had their beginnings in the development of undifferentiated power into what we know so far as the forces identified at the beginning of the chapter, and the more refined patterns such as the laws of thermodynamics. Every chance appearance of potential out of chaos—from before the point from which the Big Bang emanated—was from radical freedom. Matter—which we know from Einstein is energy—to gain differentiated being requires order and order requires consentment and limit. It also requires what, for want of more appropriate language, we can call trust as described by Simmel (1900, 1990). This ordering is the evolution of the laws of matter in concert, with power as the conductor to create the whole which is the universe. The law of everything, which, extending this metaphor might be considered to be in the mind of the composer, has so far "evaded" discovery by physicists. Notwithstanding this, the composition is awesome.

I have speculated that the telos of the universe is "being" and that the evolutionary telos is "becoming." The telos of being has, I suggest, remained steadfast. Becoming *what*, however, is determined by the composition of the evolving laws of the universe. With an understanding of Heisenberg's Uncertainty Principle, we can grasp that spontaneous, chance, and potential continue unabated out of chaos. Until or unless we find ways of discovering what happens between this confounding state of uncertainty and the reality of the universe we inhabit we are left to make a choice among an array of possible choices regarding *our* position in relation to reality. These range from radical skepticism to entrenched dogmatism. Between these extremes is the choice to trust. My "hypothesis" is that in investigating the quantum level of "mattering," we are observing the fuzzy boundary between chaos and being where potential and possibility are engaged in consentment.

To continue, Earth evolved 4.54 billion years ago, 9.23 billion years after the Big Bang (see Choi, 2017) Around a billion years (or a bit less) later, the earliest form of life evolved here on Earth. From this beginning our species evolved just 200,000 years ago. It seems to me that the evolution of life was quite the most spectacular event after the Big Bang. From those simple two-cell prokaryotes evolved: a species that not only had developed the ability to react to stimuli but one that can think. Nevertheless, as special as we may believe ourselves to be, we are not "other," *albeit* we have evolved the ability to be self-conscious, we are part of the Whole—we are equally free but, as parts of wholes, we also have limits through natural law. Consentment for us comes through our choices to accept, reject, or stay open to possibility.

Having focused, to this point, on the method of interpreting the interpretable object, as interpreter, I now follow the lead of the process physicists and step inside the process. In doing so, I am mindful that I am also the interpretable object in general as well as the self-same in particular. Along with Peirce, I am concerned not only with intelligibility but also with value. Hausman argues that value functions as a condition of intelligibility—that as he puts it (1979: 221) "intelligibility would be blind without value." Value, he argues,

> is neither a fourth category nor reducible to the categories. Instead, value must be conceived as irreducible to the categories, as co-present with them and as a cooperating condition, on equal footing with all three categories. (Hausman, 1979: 209)

Although I will be writing from here on about humans in the environment of Earth, value does not begin with us; it is co-present with the universe as it continues to create itself intelligibly.

Chapter 7

Value and Purpose

The entry for "value" in *The Oxford Companion to Philosophy* begins:

> Philosophical concern with value has focused on three connected issues: first on
> what sort of property or characteristic its "having value" or "being of value" is;
> second, on whether having value is an objective or subjective matter, whether
> value reposes in the object or is a matter of how we feel towards it; third, on
> trying to say what things have value, are valuable.
>
> These concerns closely parallel concerns with the nature of good, from which
> value is seldom carefully distinguished in philosophical discussion, though the
> terms are clearly not synonymous. (Honderich, 1995: 895)

My first concern is with clarifying the nature of "value" and with distinguishing "value" from "good" and "right."

Until the twentieth century the term used for the study of values was
"axiology" (from the Greek ἀξίᾱ, *axiā*, "value, worth"; and λόγος, logos)
but since the early 1900s "value theory" became the preferred label. In philosophy value theory is said to concern itself with the notion of goodness
and thereby with ethics and morality. The question posed by *ethics* is "What
ought I to do?" The response given by *morality* is "Live according to ethical
values." "Axiology" is the study of value or quality. "Morality" (from the
Latin *moralis*) is concerned with what it is believed is good and evil (bad) or
right and wrong. In the Western philosophical tradition value is more generally equated with *moralis* and less with *axia*. The tradition of equating value
with good, so common to the Western idiom, leads to value, good, and right
being taken as synonymous. Furthermore, as George Crowder notes:

> Monism in one form or another is the mainstream approach to ethics in the
> Western tradition stretching back to the Greeks, being a subset of the broader

philosophia perennis according to which error is many but truth, in any field, always one. (Crowder, 2003: 2)

In order to avoid what has been considered Western imperialism one could be forgiven for appealing to an idea of cultural relativism.

Cultural relativism is the view that there are no moral principles or values that apply universally, only particular moral judgments made from within the standpoint of the moral code of a specific culture. (Ibid. 7)

This notion, however, is debunked inter alia by Bernard Williams (1972: 34–39) who exposed what he labeled the "anthropologist's heresy." The epistemological mode used by the proponents of cultural relativism is usually one of phenomenology: empiricism is employed in the natural sciences and transposed to the social sciences. Isaiah Berlin, using "the technique of *Einfühlen*, or imaginatively 'feeling oneself into' the worldview of people in other times and places" (Crowder, 2003: 8), cuts through the extremes of monism and relativism with his notion of "value pluralism." The difference between what Berlin sees as value pluralism and his critic, John Gray (1995), as cultural relativism is the difference in the epistemological mode each employs—the difference cited by Mary Belenky et al. (1986: 101) between connected and separate knowing. With separate knowing "the orientation is towards impersonal rules" while "the connected mode of epistemology is toward relationship." I will discuss these epistemological modes in more detail further in this chapter, but for the moment, follow on with value.

Berlin's idea of value pluralism turns on the value of choice "according to which basic human 'goods' do not fit neatly together but are irreducibly multiple, frequently incompatible and incommensurable with one another" (Crowder, 2003: 2). He argues "the necessity of choosing between absolute claims is . . . an inescapable characteristic of the human condition" (Hardy, 2002: 214). Berlin uses "goods" in contradistinction to "evils" and the plural to underscore his idea of value pluralism.

Joseph Raz's social dependence theory presented by him as *The Practice of Value* for the 2001 Tanner Lecture on Human Values accommodates the spirit of value pluralism. He speaks of "the dependence of values on realization through valuers" (2003: 22) and of this dependence as being without relativism or reduction. His social dependence theses are:

The special social dependence thesis claims that some values exist only if there are (or were) social practices sustaining them.

The (general) social dependence thesis claims that, with some exceptions, all values depend on social practices either by being subject to the special thesis or through their dependence on values that are subject to the special thesis. (Raz, 2003: 19)

In delineating values and valuers Raz has succeeded in separating the evaluative aspect of values from the object valued, thereby breaking the nexus between values and good/right. Instead, the bond between values and valuers is realized through praxis; the relationship of values to valuers as practice is ontologically dependent, which is to say, it is dependent on the process of realization of being through a relationship.

Raz's theses pertain to society. For him, "the value of things is inert." As he sees it:

> That an object has value can have an impact on how things are in the world only through being recognized. The normal and appropriate way in which the value of things influences matters in the world is by being appreciated, that is, respected and engaged with because they are realized to be of value. (Ibid. 28)

From a Peircean perspective, no object, in being realized, is inert. Furthermore, as Peirce explained, "[t]he essence of anything lies in what it is intended to do" (CP4.659, 1908), that is, in its purpose. As I argued previously, purpose is inherent to being. Raz is aware of this but tends to express the idea in the negative, as in "there is no point to value without valuers" (p. 27). His claim that "values depend on valuers for their realization" (p. 29) comes close to my claim that without the co-presence of value, being cannot be realized. Symmetry—that valuers depend on values for their realization—requires acceptance that no object, in being realized, is inert. Despite this, Raz's ideas of practice as the relationship between values and valuers and of universal values as enabling and facilitating practice serve as a powerful model for understanding how the co-presence of value enables realization of purpose.

I will speak here to three particular values which, though commonly seen as socially dependent values—as what Raz calls enabling or facilitating values—are essential to the whole process of "mattering." Those I focus on are integrity, respect, and transparency. It is my contention that without the co-presence of values, "mattering" cannot be realized. According to this argument, the three discussed as enabling or facilitating values are categorical imperatives, that is, there are no "ifs" or "buts" about them. These are not the only facilitating values but are the ones I consider here.

Beginning with "integrity": as a socially dependent value integrity is moral "righteousness." As an enabling or facilitating value it is defined as the state of being whole and undivided. As Damian Cox, in his Stanford Encyclopedia of Philosophy entry, *Integrity* (2013), notes:

One may speak of the integrity of a wilderness region or an ecosystem, a computerized database, a defense system, a work of art and so on. When it is applied to objects, integrity refers to the wholeness, intactness or purity of a thing. (Cox, 2013)

Taking primary concern not as a fixed point, "creator," but as process, "creation" and "being" as purpose, then integrity is imperative to the process of realizing purpose. To say "beings" are rational is to say they are intelligible. By the same token, to say "being" is rational is to say it is intelligible: "being" as object is a purposeful process. Integrity, in relation to object as purposeful process, when defined as the state of being whole and undivided, is predicated on the idea of the continuum, on *synechism* as methodology and first principle. Integrity is imperative for "being" to retain intelligibility.

"Respect," the next enabling or facilitating value, requires an understanding of the relationship between "whole" and "parts." Peirce gives thirty-five definitions of "Whole (and parts)" in *Baldwin's Dictionary of Psychology and Philosophy* (1901–1905) as follows:

We may say that a whole is an *ens rationis* whose being consists in the copulate being of certain other things, either not *entia rationis* or not so much so as the whole; so that a whole is analogous to a collection, which is, in fact, a special kind of whole. There can be no doubt that the word whole always brings before the mind the image of a collection and that we interpret the word whole by analogy with collection. The idea of a collection is itself, however, by no means an easy one to analyze. It is an *ens rationis*, abstraction, or fictitious subject (but the adjective must be understood in a broad sense, to be considered below), which is individual and by means of which we are enabled to transform universal propositions into singular propositions. . . . It very often happens that an object given indirect perception as an individual is, on closer scrutiny, seen to be identifiable with a collection of parts. But it does not seem to be strictly accurate to say that the larger object of perception is identical with that abstraction, the collection of the smaller objects. It is rather something perceived which agrees in its relations with the abstraction so well that, for convenience, it is regarded as the same thing. No doubt the parts of a perceived object are virtually objects of consciousness in the first percept; but it is useless to try to extend logical relations to the sort of thought which antecedes the completion of the percept. By the time we conceive an object as a collection, we conceive that the first reality belongs to the members of the collection and that the collection itself is a mere intellectual aspect, or way of regarding these members, justified, in ordinary cases, by certain facts. We may, therefore, define a collection as a fictitious (thought) individual, whose being consists in the being of certain less fictitious individuals. (CP 6.381-383, 1901) (Baldwin, 1901–5)

He then gives definitions of the following (ibid.: 383): actual whole (James Burnett [1714–1799]), collective whole, or aggregate whole (Étienne Chauvin [1640–1725]), composite whole (Burgersdiciius [1590–1635]), comprehensive whole, constituent whole, continuous whole, copulative whole, cross whole by aggregation or aggregate whole, definite whole, definitive whole, discrete whole, essential whole (Aquinas [1225–1274]), extensive whole, formal whole, heterogeneous whole (Aquinas), homogeneous whole (Aquinas), integral whole (since Abélard [1079–1142], Blundeville [1522–1606], Burgersdiciius), integrate whole, logical whole, mathematical whole, metaphysical whole, natural whole (William Rowan Hamilton [1805–1865]), negative whole, physical whole, positive whole, potential whole, potestative whole (Aquinas), predictive whole, quantitative whole, similar whole, subject whole, subjective whole, substantial whole, universal whole, whole by accident, whole by information, whole by inherence and finally, whole by itself or per se.

"Respect" is the value commonly associated with the "Categorical Imperative" central to Kant's argument for the existence of God, where God is the primary concern—that which matters above all else—the creator and the authority of all. In Joseph Kupfer's (2016) words:

> The . . . Categorical Imperative commands us to treat all rational beings, including ourselves, with respect. . . . [T]reating someone with respect involves treating that person always as an end-in-himself and never merely as a means to our end. . . . As an end-in-himself, a person is a free, rational chooser. . . . Treating someone with respect has two dimensions.
>
> The first dimension . . . involves not *interfering* or *restricting* her as a free, rational chooser.
>
> We are permitted, however, to treat people as a means to our ends so long as we also treat them as ends-in-themselves at the same time.
>
> The second dimension . . . involves furthering her as a free rational chooser. This requires more than merely not interfering with the person as an end-in-herself; it requires actually doing something to promote the individual's free choice. https://iastate.app.box.com/v/jkupfer-phil230

Respect is imperative to the developmental, evolutionary telos of being which, though manifestly infinitely diverse and multiple, is whole. Peirce's last definition of "whole" (parts) highlights this: "Whole by itself or *per se*: a whole which essentially belongs to its parts or its parts to it" (CP6.383, 1901).

The last value, "transparency," is most commonly associated with human organizations and institutions. The nongovernment agency, Transparency International, that monitors and publicizes corporate and political corruption in international development, defines corruption as the abuse of entrusted power for private gain which eventually hurts everyone who depends on the integrity of people in positions of authority. From this it would follow that transparency means without corruption. *Tychism*, however, rules out this meaning as universal because it excludes evolutionary processes such as mutation in DNA. Defining transparency as evident and intelligible doesn't quite capture the meaning either, but it is closer.

Any concepts of democracy and society are metaphysical and reaching their meaning entails philosophical inquiry. When the concept society is skewed or simply denied reality, it is seen only as collections of individuals and not as wholes. This position has far-reaching consequences. Appeals to values such as sovereignty, nationalism, privacy, ownership, and so on give rise to conflict when these values are deemed to have been breached. In consideration of transparency, let me compare two Western societies, democratically elected, neo-liberal leaders.

Margaret Thatcher, prime minister of the UK between 1979 and 1990, said that there is no such thing as society—there were only collections of individual people. She wanted to make Britain (a place) great again where great was a measure of wealth and standing. The daughter of a grocer, she had risen to become the leader of the UK. She judged people by her own standards. During her "reign," the GDP of Great Britain (UK) did increase but vast parts of northern England were deserted with whole streets of houses boarded up and shops, businesses, and factories closed. Many of those who remained did so because they were too poor to do otherwise. I witnessed this and it was deeply distressing. I also watched with horror the increase in racism and violence. To add insult to injury, the human casualties of her politics were blamed for not bringing themselves up to her standards and the provision of welfare as a prop for the lazy. There was a period when the freedom of the individual was championed but equality was seen as something for the individual to achieve. In that she acclaimed her belief that "there is no such thing as society," Thatcher's leadership was transparent—the consequences played out as could be expected.

As far as I can remember, "society" was not in Donald Trump's lexicon, even given his intransigent outbursts about "socialism"; his concern was with numbers—with individuals. He proclaimed that he would make the United States (a place) great again. Quite aside from his penchant for lying, his blatant denial of climate change and his concomitant actions had already had a consequence on the United States during his first term; it could only be further degraded in a second term. His campaign promises lacked transparency. Joe Biden has promised to serve all Americans, even those who didn't vote for him. I hope that for their sake, he can and does.

Peirce's argument would differ. For him:

> The gospel of Christ says that progress comes from every individual merging his individuality in sympathy with his neighbors. On the other side, the conviction

of the nineteenth century is that progress takes place by virtue of every individual's striving for himself with all his might and trampling his neighbor under foot whenever he gets a chance to do so. This may accurately be called the Gospel of Greed. (Peirce, 1893)

When Peirce wrote this of the Gospel of Greed he was speaking as a nineteenth-century, white male–encultured Christian witness to the practice of colonialism, and what survived of an imperialist mindset which condoned such practices.

As the *antithesis* of agapasm, I cannot agree with his Gospel of Greed. I can only agree with his observation in terms of "every individual's striving for himself with all his might" but not to the qualification of "trampling his neighbor under foot whenever he gets a chance to do so." Freedom and equality will always be in tension, one will always qualify the other in reality. This is also noticeably so in terms of competition and cooperation, the individual and society.

I cannot believe that all the over 72 million US citizens who voted for Donald Trump in the 2020 presidential election are necessarily supporters of the Gospel of Greed. Neither can I believe that all the more than 81 million who voted for Biden did so for the greater good. I can believe, however, that Republicans favor individual freedom over concepts of equality and Democrats, as individual voters who, at the same time, appear to recognize the reality of society, tend to the reverse.

It would do well here to discuss very briefly the concept of "free will" and its relationship to freedom as boundless. From the perspective of *synechism* as the method of evolutionary development of telos, free will, although it emerged *tychisticly*, is not free of the continuum from which it evolved. Whereas freedom *qua* freedom is boundless, free will is temporal and dependent—it is embodied. Although it enables (some) freedom of purpose and choice, it is circumscribed by its embodiment on which it is dependent. That being said, I am not ignoring the enormous achievements of humans as the embodiment of free will—civilization may be considered a testament to that—but nor am I ignorant of the enormity—even the downright stupidity—of some of our choices and our co-present values.

Turning back then to the practice of values by valuers and socially dependent values.

The value most commonly associated with and expected of scientists is objectivity. Society (or an appreciable part thereof) values the work of scientists to the extent that the scientists can demonstrate their findings have been objectively produced. The main focus of Western "society" is, however, on the individual or collections of individuals, called here as organizations. By "virtue" of human law, any legally recognized organization is deemed a person. If a person or persons who are part of an organization act in ways that are in conflict with the purpose of that organization, their actions are said to lack transparency but so too can they be said to lack other values.

Here society is the valuer and its values are socially dependent. Respect, integrity, and transparency are three of the values, the co-presence of which is imperative where telos is adhesiveness. Society can adopt socially dependent values, the practice of which can and often do contravene the three mentioned values. Maybe this is what we really mean when we speak of "trouble in Paradise."

If, then, in speaking of values, one is referring to enabling or facilitating values, such values being universal are not specific and make any further imperative a *non-sequitur*. If, however, values here are taken to mean socially dependent values, then the response to any imperative can only be not necessarily.

By way of example, let me begin by discussing the practices of corporations which by "virtue" of law are deemed persons in the United States (entities in Australia) and as such can, in a limited sense, be considered as valuers. All activities of corporations as social practices are reflections of values. The practices considered acceptable are determined by external legislation/regulation and the dictates of the entities' corporate governance leaders. Of these, only the corporate governance leaders can attend to imperatives expressed by *should* in relation to activities considered appropriate and the values expressed by those activities. The values of a corporation, in view of the definition of a corporation—taken here to be a publicly listed (that is, listed on stock exchanges) private sector organization—are driven by the profit motive, regardless of whether or not these are codified in any statement of mission, vision, purpose, goals, objectives, and so on. In the case of any company listed on stock exchanges, "profit maximization" is more properly expressed as "achieving maximum return on its ordinary shares." The effectiveness of a corporation in maximizing the outcomes of this drive is dependent on the degree that all other values, as expressed in practice, can be harnessed to the driving force of the profit motive or value.

Taking a particular corporation to exemplify this: If McDonald's, a fast-food chain incorporated in the United States, is considering whether to operate in India, whose values should guide its operations there—America's or India's? Both the United States and India are democracies, are based on the English common law tradition, and as members of the WTO are market economies. While India has laws in place governing foreign investment, none of these would act as a bar to the establishment of McDonald's outlets.

McDonald's derives a high proportion of its profit worldwide through the sale of hamburgers. The meat used in McDonald hamburgers is beef. A large percentage of the American population value beef as a food source.

When it was originally established, McDonald's governance dictated that, regardless of where in the world it operated, its menu would be the same. As a corporation, its primary value is profit maximization, yet in choosing to restrict its menu choices to a demand that was determined in the United States, it restricted its appeal in the rest of the world to those countries that either shared or could be persuaded to share these particular American values. Around 80 percent of India's population is Hindus. Hindus value living cattle, not dead and served in a bun. Their choice not to eat beef is driven by values, not taste. To operate in India and meet its profit-maximizing mandate as a corporation, McDonald's needed to make a choice:

1. Set up in an area of India that is not primarily populated by Hindus.
2. Persuade Hindus to abandon their taboo on eating beef.
3. Rule out establishing outlets in India.
4. Change or expand the menu to meet fast-food demand in other parts of the world.

McDonald's chose the last-mentioned alternative and continues to do so.

As with any such corporation, the only *categorical* imperative ("do A") driving it is "the profit motive." Notwithstanding legislation and regulation, other practices and the values reflected by those activities are governed by the hypothetical imperative ("do A in order to achieve B"). Whether these values are considered good or right in the sense of ethical or moral are a sufficient but not a necessary reason for choosing to practice them. Only where profitability is impacted would a corporation that is acting within the legal/regulatory framework be *compelled* to change its practices to reflect what might be considered "good" or "right" by all or any of its stakeholders in particular and public opinion in general.

It can be seen here the co-presence of value and Peirce's normative sciences in operation. By definition, the *summum bonum* of any corporation or publicly listed company—any company listed on stock exchanges—is to achieve maximum return on its ordinary shares. This is its purpose; as its *summum bonum* it is its ultimate "good"—that which matters most. But is it "good"? For the Occupy Wall Street (OWS) protestors, whose slogan is "we are the 99%," the answer is "no." Without wading into the debate, by which I mean here, without considering motives or interests, I can say first, that what we have here is a conflict of purpose. Second, "good" here has two different meanings. The first is an end; the second, a socially dependent value. This difference can be made clearer by speaking of "good" here as "with integrity." For any corporation to say that it operates with integrity, then minimally, everything that it does must be in accord with its purpose, that is, it must be operating *effectively*. This, however, must be done within the law, not because it is the "good" or "right" thing to do, but because it is the law—albeit manmade law—that grants it its "personhood" and the law which can remove such status if it is found that it is not complying with the law. Nevertheless, to say that it is operating lawfully does not mean that it is operating with integrity in the common sense meaning of integrity as moral "righteousness," that is, *ethically*.

Shareholders who take action against a corporation that they deem is not operating in accord with its primary purpose as a corporation, may be calling its integrity, in the first sense (its *effectiveness*), into question. Alternatively, they may agree that though it is meeting its purpose, it is not doing so *efficiently*, that is, not optimally. They can vote at a corporation's AGM to demand, say, that the Chair of the Board stand down and, if they have the numbers, succeed. Unfortunately, history shows that shareholders do not tend to call corporate directors to account in the same manner for what may be seen as unethical behavior. If a corporation is deemed to have acted *unlawfully*, a class action—or group of dissatisfied customers—can argue that said corporation is breaking the law but cannot use law to challenge its integrity in terms of its primary purpose not even in the second sense of integrity, regardless of how undesirable or unethical its behavior is deemed to be. In turn, a Stock Exchange may disallow trading by any listed corporation that contravenes its own rules. Barring these exceptions, it is the buying and selling by the market that determines value. This is the economist Adam Smith's [1723–1790] model, as opposed to his predecessor and contemporary, philosopher Francis Hutcheson [1694–1746] formulator of the greatest happiness principle. Smith proposed the economic theory that social goods are maximized when individual human beings are permitted to pursue their own interests, restricted only by the most general principles of justice.

It is the sustained and extreme breach of integrity in the second sense (*ethically*) that OWS is protesting against and is directed at not only corporations but also at the likes of hedge fund managers. While corruption can be and is dealt with in law, the same cannot be said for greed which is a value that is not only being condoned but, in the not-so-distant past of the 1980s, was celebrated. OWS's purpose is to communicate the effect of rampant profit-making. In these terms it is also calling into question respect as a socially dependent value, and to some extent as a universal value.

Because epistemology is intertwined here with value, rather than ontology, let me discuss this a little more by reference to the research project published as *Women's Ways of Knowing* by Mary Belenky et al. (1986). Lest one, by reference to the title, takes this as having a female bias, the authors have recognized that the five ways are not necessarily fixed, exhaustive or universal; that they are abstract and do not account for complexity; that similar categories can describe men's ways of knowing; and that inquiry into knowing can be organized differently. In their study the authors grouped women's perspectives on knowing into five major epistemological categories as follows:

> *silence*, a position in which women experience themselves as mindless and voiceless and subject to the whims of external authority; *received knowledge*, a perspective from which women conceive of themselves as capable of receiving, even reproducing, knowledge from the all-knowing external authorities but not capable of creating knowledge on their own; *subjective knowledge*, a perspective from which truth and knowledge are conceived of as personal, private and subjectively known or intuited; *procedural knowledge*, a position in which women are invested in learning and applying objective procedures for obtaining and communicating knowledge; and *constructed knowledge*, a position in which women view all knowledge as contextual, experience themselves as creators of knowledge and value both subjective and objective strategies for knowing. (Belenky, 1986: 15)

Following their initial discussion, the authors moved beyond this scheme and referring to Carol Gilligan's *In a Different Voice* (1982) adopted her terms, *separate* and *connected* knowing. Gilligan's study was intent on exposing the deficiencies in Kohlberg's model of moral development. Kohlberg had conducted his research both longitudinally and latitudinally, but his subjects were all male. Gilligan conducted the same research but with female subjects.

> When one begins with the study of women and derives developmental constructs from their lives, the outline of a moral conception different from that describe by Freud, Piaget, or Kohlberg begins to emerge and informs a different description of development. In this conception, the moral problem arises from conflicting responsibilities rather than from competing rights and requires for its resolution, a mode of thinking that is contextual and narrative rather than formal and abstract. (Gilligan, 1982: 19)

As she says, "The morality of rights differs from the morality of responsibility in its emphasis on separation rather than connection, in its consideration of the individual rather than the relationship as primary" (p. 2). Gilligan stressed that while her empirical observations were of women,

> this association is not absolute and the contrasts between male and female voices are presented to highlight a distinction between two modes of thought and to focus on problems of interpretation rather than to represent a generalization about either sex. (Ibid.)

In the literature, these and other similar studies have largely disappeared, like water poured on desert sand, with separate knowing remaining the hegemonic mode employed. Yet, when only strictly objective critical analysis is recognized as valid, polemics and dualism result, and as Belenky et al. noted:

> Presented with a proposition, separate knowers immediately look for something wrong—a loophole, a factual error, a logical contradiction, the omission of contrary evidence. (Belenky, 1986: 104)

The implicit value is in competing, in being right, in winning. The promise of certainty, however, is illusory. Moreover, it is a dangerous game to play in the social sciences. It makes objects of subjects. It lacks respect. In the context of the subject of their study, that is, of women's ways of knowing, Belenky et al. point out that in relation to procedural knowledge as defined earlier:

> Women who rely on procedural knowledge are systematic thinkers in more than one sense of the term. Their thinking is encapsulated within systems. They can criticize a system, but only in the system's terms, only according to the system's standards. Women at this position may be liberals or conservatives, but they cannot be radicals. If, for example, they are feminists, they want equal opportunity for women within the capitalistic structure; they do not question the premises of the structure. When these women speak of "beating the system," they do not mean violating its expectations but rather exceeding them. (Ibid. 127)

Returning to Honderich's entry for "value," where he records that the second philosophical concern with the issue is of "whether having value is an objective or subjective matter" (1995: 895). This notion of objectivity hides a multitude of "sins," not least of which is that it is value laden. Once we understand that objectivity is not necessarily "right" we can see it as such and looking back to the Ancient Greek philosophy, as the position promulgated by the Stoics as *apatheia* (without passion). The value of *ataraxia* (tranquility) was also recognized by the Stoics but for the Epicureans it was of primary importance. I am not well-read enough in ancient history to know how the Stoics and Epicureans managed their differences, but I would hazard a guess that the latter believed theirs the "good" life and the former theirs the "right" one.

Harking back to Peirce's contention that "[t]he essence of anything lies in what it is intended to do" (CP4.659, 1908), that is, its purpose: objectivity, for him, included mathematics, which he was discussing when he made this statement and which he insisted be conducted diagrammatically. Susanna Marietti (2010: 149) points out "[a]ccording to Peirce, a diagram is a sign that makes the relations between the objects represented perceptible." Claudine

Tiercelin, in her essay "Peirce on mathematical objects and mathematical objectivity" points out that

> he thought that the *meaning* of mathematical statements could not be given independently of any demonstration; in that sense, although Peirce's pragmatistic realism about indeterminacy prevented him from reducing the meaning of a proposition to its conditions of verification, or reducing meaning to use, Peirce never separated the meaning of any mathematical proposition from its conditions of *assertibility*. (Tiercelin, 2010: 105)

This could imply that Peirce was an anti-Platonist—usually meaning *anti-realist*—but this would only be so if one viewed such diagrams as particular rather than general; as static rather than processual. Take, for example, what Frank Wilczek (2008: 19–20) calls "Einstein's second law," $m = E/c^2$. This is a manipulation of $E = mc^2$ which hypothesized that a body's mass arises from the energy of the stuff it contains, and which suggests the possibility of getting large amounts of energy from small amounts of mass. In this form it calls to mind nuclear reactors and nuclear bombs, which is what hippies, such as myself, protested against. Manipulating the "diagram," however, as in, $m = E/c^2$ suggests the possibility of explaining how mass arises from energy, and it was this turn that led to the search for the Higgs boson. Study of the experimental data of the Higgs, which was isolated by the LHC mid-2012, has suggested, however, that it is not "the One."

When still operating, the Process Physics team at Flinders showed that valuing objectivity in science to the exclusion of the knower, the interpreter, stymies discovery. The value of what Belenky et al. have termed "constructive knowing" and Gilligan (1982) as "connected knowing," was being realized. Kirsty Kitto whose PhD thesis *Modelling and Generating Complex Emergent Behaviour* (2006) says of this shift in value:

> Through a proper treatment of context we lose much of the confusion that often surrounds an examination of the Universe; an observer of the Universe can exist *within* that Universe. (Kitto, 2006: 10)

Those scientists who do not recognize objectivity as value laden—who are in its thrall—are unwittingly circumscribing their purpose as scientists. While their integrity, in both senses, may remain intact, they may be forced into a position of compromising other values or their purpose. This compromising position, commonly called "conflict of interest," might better be described as "conflict of purpose." Whichever, it lacks transparency.

There is no absolute good or bad, right or wrong—no certainty. There is only better or worse. Values are not abstractions—valuers value, and values are a property of a relationship between valuers and that which they value. At the heart of value is contingency. What this suggests is that values are intentional, that is, that they are purposeful, which in turn means they are teleological.

Demonstrating this calls for a more complex example than the one given earlier of McDonald's. I have chosen to consider the conduct of clinical trials here and in chapter 8. I will speak of the public sector organizations that conduct clinical trials—most often universities—as Clinical Trials Centers (CTCs) and those operating in the private sector as Contract Research Organizations (CROs).

Despite their research status, the bulk of trials CTCs conduct is likely to be industry sponsored. Centers seeking funding for trials that are in the public interest but which are unable to demonstrate that such trials would "add value" to the primary purpose of a funder are often engaged in protracted search for funding. An example of such a trial type was the Benefit of Oxygen Saturation Targeting (BOOST) trial concerned with achieving adequate delivery of oxygen to premature babies without creating oxygen toxicity. The trial was funded by the National Health and Medical Research Council (NHMRC) in Australia, the New Zealand Health Research Council (NZHRC) in New Zealand, and the National Research Council (NRC) in the United States. The data from these BOOST trials was then pooled with the data from Surfactant, Positive Pressure, and Pulse Oximetry Randomized Trial (SUPPORT) which was funded by the National Institutes of Health (NIH). The "routine" use of supplemental oxygen in the care of preterm infants originated from observations in the 1940s leading to the widespread practice of unrestricted oxygen supplementation for small or sick infants. Unfortunately, though survival increased, so too did severe eye disease and blindness. Kate Campbell from Australia suggested in 1951 that oxygen could be responsible for this rising epidemic. It took more than fifty years (from 1951 to 2003 when SUPPORT was launched) to garner funding for large randomized clinical trials to investigate Campbell's hypothesis.

As I argued earlier, all activities of corporations, as social practices, are reflections of the values of their owners. The practices considered acceptable are determined by external legislation and regulation and the dictates of entities' corporate governance leaders. The values of corporations, in view of the definition of a corporation, are driven by the profit motive. The effectiveness of maximizing the outcomes of this drive is dependent on the degree that all other values, as expressed in practice, can be harnessed to the driving force of this motive or value, which is its primary purpose. Only where profitability is impacted would corporations that are acting within the legal and regulatory framework, be compelled to change their practices.

Harnessing the public sector and its values, in the pursuit of maximizing shareholder value, is in the interests of big business. Alliances between the sectors are in the interests of national economies on the one hand and profit maximization on the other. This is very much so in the case of the pharmaceutical industry whose return on investment (ROI) has been in double figures for the multinationals. Graham Dukes of the Unit of Drug Policy Studies at the University of Oslo notes, however, when considering the World Health Organization's report on Priority Medicines for Europe and the World:

> Public-Private partnerships can lead to a confusion of priorities and interests. The public P and the private P do largely have different purposes and interests, however much one tries to harness them together and if they are harnessed together there is a risk of the one dragging the other onto its own course. (Dukes, 2004)

In the case of public research, the problem is exacerbated by governments' insistence on universities and other public research organizations supplementing their government research income with industry funding, often making securing of such buy-in a prerequisite for receiving government grants. Cutting the funding of public entities has, in many instances, put them in a position where survival of research programs is contingent on industry alliances. This, in turn, has put governments, their instrumentalities, and their staff at enormous risk of entrenched conflict of purpose. To counter this endemic situation, it is necessary to understand first, how, despite acting technically within the law, corporations have succeeded in convincing governments to give their interest precedence, even when this is manifestly not in the public interest. Michelle Brill-Edwards of the Health Protection Branch of Canada's Centre for Health Services and Policy Research (2000) is very aware of the heuristics of this. The first tactic, she says, is to deny for as long as possible that there is an issue through the strategy of divide (and rule). This is achieved by discrediting any complaint that is not in the industry's interest. Such tactics include labeling detractors as "biased," "unreliable," "disgruntled" "impractical," "the loony fringe," "difficult to work with," and so on. If a situation puts the issue in the wider public spotlight, delay tactics often provide an almost foolproof way of getting it off the public's radar long enough to regroup.

Ultimately, the industry will accede to legislation when flagrant disregard can be established. It will do so to protect and retain its gains achieved through dismantling the public sector's stronghold on interests that are in conflict with the profit motive—gains achieved by the pharmaceutical industry largely through, what Brill-Edwards describes as, a protracted process of dismantling the regulatory structures with which drug approval has been governed. This process, which has involved three separate but related tracks, de-regulation, de-professionalization, and de-construction, appears engaged in removing anything in the way of getting new products to market in the shortest time possible. FDA regulation is now paid for by fees from the industry, a situation that Dorey (2004) notes "has led to the industry becoming treated as the customers, and regulators being anxious to serve the industry." A particularly relevant outcome of this reversal has been the introduction of what is called: "Notice of Compliance with Conditions." This permits a company to market drugs without prior evidence of efficacy or safety. All that is required is evidence described as "promising." The only condition on such marketing is that the company commits to carrying out more research at some (unspecified) later date. This is an inversion of the previous standard of burden of proof required to market a drug—now causal proof of harm needs to be produced before consideration is given to its withdrawal (Brill-Edwards, 2000). Lexchin (2005) quotes a prediction by Drummond Rennie, editor of *Journal of the American Medical Association* (JAMA) that dependence of clinical research on money from the pharmaceutical industry will change only "when commercial ties are linked to deaths" (in Greenberg, 2003, Lancet 362: 302-303). The remedy, instituted by the US Department of Health and Human Services' National Institutes of Health (NIH), is to clamp down harder on individuals.

> Under the new rules, all NIH employees are prohibited from engaging in certain outside employment with: (1) substantially affected organizations, including pharmaceutical and biotechnology companies; (2) supported research institutions, including NIH grantees; (3) health care providers and insurers; and (4) related trade, professional or similar associations. Investments in organizations substantially affected by the NIH, such as the biotechnology and pharmaceutical industries, are also not allowed. (February 2005)

This move, which provides no compensatory measures, is playing right into the hands of an industry intent on de-professionalizing the public sector by stripping it of capable professionals—physicians, scientists, chemists—who can intelligently evaluate data and have the professional capacity to defend their decisions. Even given that the FDA "submitted to a far-reaching reform of its drug approval procedures following the Vioxx debacle" (Dorey, 2004), there has been no suggestion that "Notice of Compliance with Conditions" will be rescinded or that other, more systemic conflict of purpose will be addressed.

The increasingly dependent relationship between government and industry represents a shift by governments, from the public interest to the interest of the industry. Such alliances have created entrenched, systematic conflicts of purpose which only a systematic solution can address. It is not up to the industry—nor is it in its interest—to provide the solution; the industry has no conflict of interest. Neither should the onus of responsibility continue to fall on individuals alone; this is a response that focuses on effects and not on causation. Rather it is the responsibility of governments to honor the social contract and to establish systematic ethical practices for bringing about a more equal contest between public and private interests. Such a move is unlikely to occur under government authority that does not recognize society.

This process, with its conflict of purpose, lacks transparency. It could be further argued that it also lacks respect and integrity. At an international level, the response to the Covid-19 pandemic is revealing these values or lack thereof.

Chapter 8

Value and Power

I said at the outset, quoting Mary Gaudron: "The matters that matter may differ depending on who is doing the mattering."

In chapter 7 I showed that "mattering" is grounded in and powered by value and purpose, that is, I focused on the values and purpose of *who* (or *what*, I must add) is doing the "mattering." In this chapter I focus on the *doing* of "mattering." In Peircean terms, this is the category of Secondness of my hypothesis: Power—where power is the capacity to cause—is the enabler of force functioning as actual "mattering."

Because power is as ubiquitous as value and because in reality "mattering" is irreducible, of necessity, in discussing value, I have made reference to power in the previous paragraph. Here, however, I show that power enables the "mattering" of the matters that matter expressed through purpose and values to produce consequences along continuums of probability.

I discuss the utilization of power by human society (I assume the reality of society) and specifically as it relates to paid employment. In doing so I compare the organizational models known as "Best Practice" Model and "Best Fit" Model as a means of assessing which is best aligned with society's needs. In the first instance I describe the two models but know that, even though I do not show this level of detail, the models need to be translated into flowcharts and networks. Like Peirce's existential graphs, this method makes observable the logic of processes and the relationships of the parts to the whole organizations as they systematically evolve within an environment. Flowcharts and networks are diagrammatical representations of algorithms. Together with other tools for assessing quality such as histograms, Pareto charts, check sheets, control charts, cause-and-effect diagrams and scatter diagrams, flowcharts and networks reveal the operation of the algorithms—that is, their theoretical power.

The choice of either the "Best Practice" Model or the "Best Fit" Model as strategy is determined by the purpose and the values driving the choice of such purpose. The choice of model, regardless of its function, has differing effects on the people—the valuers—that make up organizations.

First, a description of the models, particularly in relation to what organizations of all types call human resource management (HRM).

The most commonly recognized HRM "Best Practice" Model was developed by Jeffery Pfeffer (1998) and whittled down from sixteen to seven dimensions:

1. employment security;
2. selective hiring of new personnel;
3. self-managed teams and decentralization of decision-making as the basic principles of organizational design;
4. comparatively high compensation contingent on organizational performance;
5. extensive training;
6. reduced status distinctions and barriers; and
7. extensive sharing of financial and performance information throughout the organization.

Pfeffer, though agreeing that several of these dimensions "appear to fly in the face of conventional wisdom," explained their underlying logic and argued

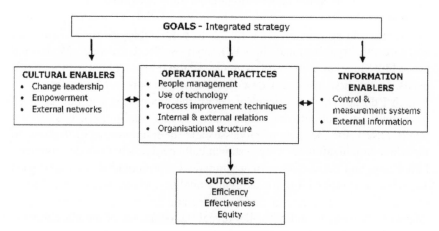

VISION: *the leader in measures of excellence, the operational benchmark and the preferred organisation in the particular market.*

MISSION: *to achieve organisational* **BEST PRACTICE**

Figure 8.1   A Best Practice Model. *Source*: Adapted by Sophia, 2021.

that for success in achieving high-performance organizational arrangements, it is counterproductive to take this approach piecemeal (Pfeffer, 1998: 96–124). The "Best Practice" Model shows as arising from an organization's vision and following from its vision and mission as shown in Figure 8.1 (above).

This "Best Practice" Model is necessarily transparent—choosing the model means choosing transparency. Not so the "Best Fit" Model (depicted next) which only discloses that which is required by law and what it *chooses* to disclose. It is correct to say that while "Best Fit" may be systematic, as suggested by the beehive structure depicted further, it is a closed system, not only externally but internally also.

In terms of its purpose, an organization opting for "Best Fit" will always optimize effectiveness and efficiency but not so equity. Freedom and equality may be uttered in the same breath, but any move toward the latter will always compromise the former. This can account for resistance by private enterprise to any regulation which circumscribes efficiency. Freedom is categorical while equality is a value of choice. This may be interpreted as "survival of the fittest," but such interpretation shows a lack of understanding of the evolution of ecosystems to which rampant freedom is anathema. A generic HRM "Best Fit" model which is contextual and closed as exemplified in Figure 8.2.

The operation of organizations is dependent on its value and purpose or intention. Let me exemplify this by bringing into focus roles within public and private sector clinical trials centers—clinical trials centres (CTCs) and contract research organizations (CROs). Such centers expect work to be performed in accordance with their purpose, vision, mission, values, goals, operational rules and outputs, and outcomes to result from performance. This

Figure 8.2   A Best Fit Model. *Source*: Adapted by Sophia, 2021.

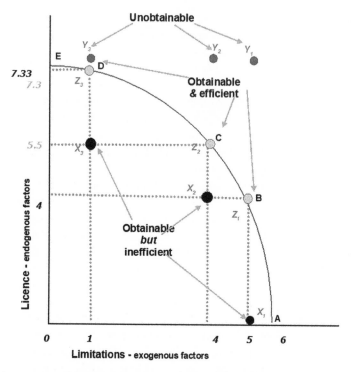

Figure 8.3   A Performance Productivity Frontier Where Performance Is Framed as Outputs. *Source*: Adapted by Sophia, 2021

is called accountability: achieving purpose is being effective; doing so with no more than is necessary to be effective is considered efficient. Performance in these terms is measurable with a model such as that shown in Figure 8.3.

Organizations dedicated to conducting clinical trials typically employ staff engaged in core activities including medical scientists, clinical trialists, statisticians, biostatisticians and support staff of computer technicians and administrators. Every employee has license, circumscribed by limitations. On average in the public sector, core activity employees have greater license but are paid less than their counterparts in the private sector. Plotting obtainable and efficient performance expectations along the productivity frontier where time is the limiting factor, a biostatistician who may be required to do research, teach, "number crunch" and do administrative tasks may plot at *B*. Increasing productivity—moving up the curve to *C*—means freeing-up time. This can be done in a number of ways: by way of overtime, by use of own time, and by rationalizing the role. Where economies of scale can be achieved, a team would share the load and may be able to operate along the *BC* curve, or even the *BD* curve. If, say, 20 percent of the workload of biostatisticians operating at *B* is taken up with teaching, then the opportunity cost of teaching within a proscribed workload is 20 percent. If, however, this is an average over a year and an academic year is eight months, then when counting only face-to-face teaching (together with student assessment), the allocated time may be reduced to 13 percent. This, however, lacks transparency if the assumption is that such time-consuming activities such as staying abreast in the field, curriculum development, lecture course planning, mentoring, and coaching are excluded from the load, or that these will be conducted in the biostatisticians' own time. This is not an issue for private sector clinical trial centers where teaching is not included in their purpose.

At this point it is worth clarifying the difference between wages and salaries. In a general way, wages are paid to employees for their time doing work; salaries are paid for getting work done. Those earning wages would (or rather, should) be paid for overtime, whereas, employees who are earning a salary are being paid for the work performed and time taken to do work in excess of time at work would be done in their own time. In Australian universities, all nonexecutive employees—both tenured and non-tenured—are paid wages. Those whose roles are primarily devoted to research and teaching are employed on the academic award with over-award remuneration being paid as loadings. Research and teaching academics do not expect, nor are they expected to be paid for overtime even though they are formally being paid for time worked. Ostensibly this is because of a presumption that academics can manage their own workloads putting more hours into teaching in semester sessions and less during the breaks. Even so, a research center may decide they are not in a position to allow for such flexibility. This is compounded when, disregarding some aspects of the overall teaching function necessary for meeting all their accountabilities, only say 5 percent for a 10 percent role is accommodated.

Regardless of whether this is an error in design or an intentional play on the expectation of academics using their own time it is a breach of integrity by the employer. It also shows a lack of respect and transparency. In this case, lack of respect expressed in terms of Kant's categorical imperative means the use of employees beyond a mutual contract as a means to its own ends. The absence of transparency is the obfuscation of the corruption of the process. The consequences of any breach of values by employers may show up in a number of observable ways: constantly "overworked" employees, high staff turnover, hostility between employers and employees, low morale, and tit-for-tat breaches of values by employees, to name but a few. Such situations are common and complex even when considered in terms described above where the attributable cause is workload expectations.

Let me put myself in a senior administrative role of a university CTC as a means of resolving such an untenable situation. Before applying the model, I would need to do a great deal of preparatory work including making a study of the clinical trials "market" and its history and the management strategies of the sponsors and funders of medicine trials; identifying the CTC's purpose, assessing whether this was aligned with the university's purpose and ascertaining the skills, abilities, experience necessary to meet the purpose and the resources, including funding required to realize its achievement; and negotiating with the CTC's academic and trials directors to develop the vision mission and goals; spelling out the consequences of including such concepts as excellence, benchmarking, and best practice. A choice for inclusion of these would suggest the operating rules. At the same time, I would need to be identifying endogenous and exogenous power sources and their predictable impact.

Beginning with the history: two phenomena deserve special attention. First, the passing of the 1980 Bayh-Dole Act, the purpose of which was to encourage the utilization of inventions produced under US federal funding. The Office of Intellectual Property Management of the University of New Hampshire, for example, issued a statement to its research community (2002) prefacing it:

> This policy promotes the participation of universities and small businesses in the development and commercialization process. It also permits exclusive licensing with transfer of an invention to the marketplace for the public good. The government gets a royalty-free, non-exclusive license to use for government purposes (including use by government contractors).
>
> The policy permits universities (all non-profits) and small businesses to elect to retain title to inventions made in performance of the federally-funded program.
>
> It was understood that stimulation of the U.S. economy would occur through the licensing of new inventions from universities to businesses that would, in turn, manufacture the resulting products in the U.S. (Office of Intellectual Property Management, University of New Hampshire, 2002)

An unfortunate but unintended consequence of the passing of the Bayh-Dole Act and the removal of the firewall between public and private funding, was a shift of clinical trials of pharmaceutical medicines by pharmaceutical companies from universities to the private sector, in particular to newly emerging clinical trials research companies such as Quintiles and Covance. The ensuing decades saw an inversion of a 4:1 ratio in universities' favor in the 1980s to that of private sector companies this century. Remembering that between 1980 and regulation clamping down on practices perceived as conflicts of interest, industry has had more than four decades during which to siphon off clinical trials expertise from universities to serve its own interests; public interest has been a casualty.

Second and not unrelated, was a change of focus by pharmaceutical companies from improving revenue to cutting costs (Kager & Mozeson

2000)—to a focus on "productivity" and "value added." This change has not been confined to the pharmaceutical industry; it has been taken up across industry and by public funding bodies including governments. Because of the "Best Fit" Model's lack of transparency—often in the name of "commercial-in-confidence"—it is only by keeping a close eye on trends and on the media that an outsider might gain insight into what has been and is, going on.

This shift of responsibility for funding public health research from the public to the private sector is a reflection of a trend by Western governments to support the economic theories of Adam Smith [1723–1790] rather than those of John Maynard Keynes [1883–1946]. I have not considered the third major modern economic theory, that of Karl Marx [1818–1883] because it has been largely abandoned by Western powers. On the one hand, John Maynard Keynes's economics is an approach to economic policy that favors using the government's power to spend, tax, and borrow to keep the economy stable and growing. On the other hand, Adam Smith advocated a *laissez-faire* attitude by the government toward the marketplace, allowing the "invisible hand" to guide everyone in their economic endeavors, theoretically creating the greatest good for the greatest number of people and generating economic growth (Gorman, 2003).

Which of these two major economic doctrines individuals tend to value will directly affect their desires, endeavors, expectations, and ethics? As noted earlier, research scientists in the private sector are in general paid higher salaries than those with comparable roles in the public sector, but— and it is a big but—their freedom or license is more circumscribed. It could be said that this translates as greater remuneration satisfaction for less job satisfaction. Most notably research scientists in the private sector forego the freedom of publication or sharing of their research findings and freedom and time for conducting "blue skies" research and for teaching. Any university CTC wanting to staunch the flow of their people to the private sector needs to be very aware of this.

With this in mind, as the senior administrator, I next need to identify the CTC's purpose and assess whether it is aligned with the university's purpose. Following this comes the task of listing the generic functions involved in operating a university CTC. These include identifying public and community health issues requiring new or novel treatments; developing hypotheses addressing these; designing appropriate trial programs; exploring novel approaches in biostatistics for interpreting data and finding meaning; conducting systematic reviews and clinical trials; teaching and supervising university undergraduate and postgraduate students; developing and presenting education and training programs and methods for translating evidence into practice; engaging in quality control and performance reviews; and finally keeping up to date in the field. In addition, everything involved in information technology, human resource services, contract development, financial and accounting services, site management, reception, and other administrative support must be accounted for. Following this, is the need to draw up the matrix of duty statements, identifying the essential job and candidate criteria and classifying each accordingly. Ascertaining the workloads and the concomitant cost and adding a percentage for non-payroll expenditure, then determines the recurrent income stream required to fund the activities of the CTC. Finally comes the task of developing budgets, making grant applications, and seeking sponsors.

The concurrent exercise is of developing a strategic plan for performance in accord with purpose and within resource constraints, focusing on a vision and mission, identifying the values necessary to meet these, specifying the goals and suggesting the operating rules. A top-level plan might look something like this:

Clinical Trials Centre (CTC) of a university

CTC's purpose is to contribute to national and international evidence-based health care in public and community health and in the acquisition, creation, analysis, interpretation, synthesis, translation and transmission of knowledge concerning clinical trials.

CTC's vision of itself is as the leader of clinical trials research excellence, the benchmark for the conduct of investigator initiated clinical trials and the preferred reference for translating clinical trials evidence into practice.

CTC's mission is to achieve best practice in health care and improve outcomes through the use of clinical trials.

CTC strives to co-operatively demonstrate its values of commitment to public and community health and education and learning, integrity, respect, transparency and rigor in all its planning, decision making, actions and interactions.

CTC's goals are to:

- generate high quality evidence of the effectiveness of health care interventions through randomized trials;
- be a national resource in design, conduct, analysis and interpretation of randomized trials;
- improve evidence-based health care through the use of clinical trials and high-quality systematic reviews of trials;
- provide high quality education and training in design, conduct, analysis and interpretation of clinical trials;
- develop methodologies for designing and undertaking relevant clinical trials and for combining and interpreting trial results for better practice;
- develop and assess strategies for improving the translation of trial evidence into best practice.

CTC's operational rules:

- How-to rules—spell out key features of how activities and processes are to be executed, as:
  - Best Practice
  - Resourced
- Boundary rules—focus on which opportunities can be pursued and which should not, i.e.:
  - In accord with the vision, mission, values and goals
  - Stretching (but not exceeding) present or acquired competences
- Priority rules—rank the accepted opportunities:
  - In order of their capacity to realize the vision or to lead to activities that can increase that capacity
- Optimizing rules—synchronize and enrich activities by:
  - Consulting, cooperating, coordinating, communicating and coaching
- Exit rules—decide when to pull out of yesterday's opportunities by:
  - Spelling-out objectives and performance indicators and conducting periodic reviews

The minimum values that must be practiced by *all* parties at all times if any organization is to be effective—that is, continually achieve its purpose in terms of its vision and mission—and efficient—that is, optimally performing in accord with its operating rules—are commitment to public and community health, integrity, transparency, respect and rigor. Other values may be added, but they, like these, must be matched with performance indicators if they are to be considered meaningful and systematic. The purpose, vision, and mission of any organization are value driven and maintained. Values coexist with performance—different values produce different outcomes, as does a lapse in such values in any part of the organization. Those values that are identified as essential must be both ubiquitous and continuous. This is so for both employer and employees and thus must operate in performance reviews—performance is grounded in values.

The above plan is cast in terms of expectations of employees but is silent on what employees may expect of the employer. Returning to the seven dimensions of the "Best Practice" Model: the employer must attend to these if there is to be symmetry of values and concomitant performance between employer and employees in the effective and efficient achievement of the organizations' purpose, vision, mission, and goals. Taking these one at a time and remembering that they are grounded in the organization's values:

1. *Employment security* for university employees means tenure. Tenure or tenure-track contingent on performance for both academic and non-academic employees must be the norm. The "Best Fit" Model rarely offers this highly valued dimension. Instead, it may offer something like a 10 percent loading for fixed-term contracts.

2. *Selective hiring of new personnel* is a commitment that can only be made as positions become vacant or with growth. Self-attrition by tenured employees is often considered the only way an organization can free itself of non-performers, but this is not the case if the employment contracts of unsatisfactory performers are terminated. It must be stressed that the reflection of the organization's values in performance is the focus of assessment. A tool for making such an assessment would first need to be designed and tested by a highly competent team: all the many dozen performance appraisal models and "kits" available on the market are woeful in my estimation.

3. *Self-managed teams and decentralization of decision-making as the basic principles of organizational design* require transparency and close attention to change management both of which first require extensive exploritative learning.

4. *Comparatively high compensation contingent on organizational performance* is not just a matter of money. Time for employees to pursue activities in their shared interest with sponsoring universities is highly valued as is control of research data and freedom of publication of results. Here the ability of CTCs to negotiate contracts with funding bodies is essential.

5. *Extensive training* not only in operational matters but also in change management, team building, conflict management, problem-solving, decision-making, and understanding values, to name a few, are essential.

6. *Reduced status distinctions and barriers* can be "the rub." Dispensing with the use of titles, for example, barely scratches the surface. More important is attention to endemic differences in status ascription in universities between academic and non-academic staff and between a medical model and what may be called a nursical model of health, where the former is directed toward a cure and the latter focuses on care. Academics are routinely considered of higher status than non-academics, as is the medical rather than the nursical model of health. Such differential status ascription is hegemonic. Putting aside, for the moment, consideration of barriers, and status differentials are always a matter of opinion grounded in value. Any attempts to resolve differences of opinion are doomed to failure, but there is an alternative approach, one put forward by Yasuhiko Genku Kimura in his paper "Alignment Beyond Agreement" (2003) in which he explains:

Alignment is congruence of intention, whereas agreement is congruence of opinion. Opinion is a supposition elevated to the status of a conclusion held to be right but not substantiated by positive proof—rational or evidential. Because disagreement means difference of opinion, disagreement often escalates into a dispute as to whose opinion is right. . . . Alignment does not require agreement as a necessary condition. Alignment as congruence of intention is congruence of resolution for the attainment of a particular aim. . . . The question is not "who is right" but "what is best" for the fulfilment of the intention. (Kimura, 2003: 1)

The intention is written into the purpose, mission, vision, and goals of the CTC and into those same for the university sponsoring the CTC.

7. *Extensive sharing of financial and performance information throughout the organization* requires complete transparency. This is not merely a matter of making data available; as information sharing, it is focused not so much on knowledge, as on understanding the reality of the event that is the organization "mattering."

This is all "well and good"—one may even say "noble"—but if, as I seem to be implying, a "Best Practice" Model is better than a "Best Fit" Model, what makes it better, and in what respect does the preferred model depend on Peirce's system.

The first thing to reiterate about the "Best Practice" Model is that to achieve its purpose *optimally* it must be practiced as a whole. As with Peirce's method, it is temporal, is future directed, its growth and development are evolutionary, and it is mediated. To say it is mediated is to say it is attended to by itself for its well-being and flourishing as a whole within an environment. Power as attention is practiced throughout the "Best Practice" organization and not merely at the top. I use the term "attention" here as the mediator rather than Peirce's "agape" because, you might say, the former is better understood during "business hours." Attention, ramped up a notch to include commitment comes closer to what Peirce meant by "agape." Whatever the term, its meaning remains as does its necessity as mediating power.

As organizations, both models are "mattering" and, if my hypothesis is valid, as "mattering," both models are grounded in values and are purposeful; are actualized through the function of causation; and are realized through evolutionary growth and development. The "Best Practice" Model assumes that a set of practices aimed at high commitment and high performance will benefit all organizations regardless of context. Taking "fit" in "Best Fit" to mean possessing or conferring the ability to survive in a particular environment, then, as with Herbert Spencer's biological evolutionary theory of "survival of the fittest," the model is predicated on an assumption that survival is primary. "Fit" is concerned externally with competitive strategy and internally with coherence. In the jargon of HRM the "mattering" of the "Best Practice" Model is "soft," stressing the "human" aspects; the "mattering" of the "Best Fit" Model is "hard" and is focused on high marginal economic returns.

Deriving from their purpose for "mattering," both models are concerned with return on investment. In the case of private sector organizations, this return is profit, regardless of whether or not this is made explicit or is merely implicit in being private sector organizations—it is a commitment by the "owners" of the organization to achieving the highest profit. Return in the public and Not-For-Profit (NFP) sectors is the outcome of commitment by the parts to the whole.

Regardless of the model utilized, living people are the "mattering": they are the market, the inputters and outputters, the suppliers and the demanders. For people to matter—that is, to live—needs must. The Hierarchy of Needs developed by Abraham Maslow [1908–1970] illustrates these. Consideration of an *organization*'s responsibility for meeting each of these levels is a telling exercise. In order of need they are Physiological, Safety and Security, Belongingness, Esteem, and Actualization.

First, though, so as not to create confusion and to accommodate understanding, I need to refer to the concept of "wholes (parts)" raised in chapter 7. Even though Maslow's model is concerned with individuals, it makes two assumptions—an environment (a context) and other individuals—necessary for meeting the needs of the individual. Individual people when viewed minimally as engaged in maintaining homeostasis are organizations; in and of themselves they are wholes. This can be interpreted as what Peirce meant when he claimed that the individual is not real. The real—"mattering"—is irreducible beyond First, Second, and Third. Furthermore, as John Donne the English metaphysic poet wrote in his *Meditation XVII* (1624):

> No man is an *Iland*, intire of it selfe; every man is a peece of the *Continent*, a part of the *maine*; if a *Clod* bee washed away by the *Sea*, *Europe* is the lesse, as well as if a *Promontorie* were, as well as if a *Mannor* of thy *friends* or of *thine owne* were; any mans *death* diminishes *me*, because I am involved in *Mankinde*. (Donne, 1624) [Donne's original spelling and punctuation]

Not to diminish self-responsibility given persons are whole organizations, human society as the whole organization of its parts—people—is responsible for meeting the needs of its parts. Society, however, is spread (unevenly) around this world (environment) which other organizations and their composite parts—people—inhabit. Society as a whole evolves as an "organization" of "organizations" of people such as nations, "races," cultures, religions, and markets. These "organizations" are all the creation of people as are their "borders." Here the claim of sovereignty is highly pertinent. A claim of sovereignty is a claim of supreme controlling power and absolute and independent authority by a person or persons, or in cases of claims on behalf of transcendent power/s (e.g., God/s), spokespersons through scriptures.

Markets, however, can and do operate over sovereign boundaries. As discussed in the previous chapter, global corporations/organizations are considered as persons, but unlike human persons, they can cross national borders unimpeded.

All this is pertinent to value, power, and realization of "mattering." Let me now return to consideration of an organization's responsibility for meeting each level of Maslow's model of Hierarchy of Needs and whether such consideration is relevant to an interpretation of the "Best Practice" Model as better than the "Best Fit" Model.

The five levels of the model are:

1. *Physiological* needs which are the physical requirements for human survival. If these requirements are not met, the human body cannot function properly and will ultimately fail.
2. *Safety and Security* needs which include shelter personal security, financial security, health and well-being, safety nets against accidents/illness and their adverse impacts. In the absence of economic safety—due to economic crisis and lack of work opportunities—these safety needs manifest themselves in ways such as a preference for job security, grievance procedures for protection from unilateral authority, savings accounts, insurance policies, reasonable disability accommodations, aged care, and so on.
3. *Belongingness* which is interpersonal. According to Maslow, humans need to feel a sense of belonging and acceptance among their social groups, regardless if these groups are large or small.
4. *Esteem* which concerns the desire to be accepted and valued by others. Maslow noted two versions of esteem needs: a "lower" version and a "higher" version. The "lower" version of esteem is the need for respect from others. The "higher" version manifests itself as the need for self-respect. This "higher" version takes precedence over the "lower" version because it relies on an inner competence established through experience.
5. *Actualization* which refers to what a person's full potential is, and the realization of that potential. Maslow describes this level as the desire to accomplish everything that one can, to become the most that one can be. As previously mentioned, Maslow believed that to understand this level of need, the person must not only achieve the previous needs but master them.

Maslow, in his later career, came to see that the "hierarchies are interrelated rather than sharply separated." His model is still widely used across a number of disciplines but is not without its critics. Nevertheless, such criticisms, once identified, can be largely overcome and if not, then its shortcomings are recognized. Putting aside criticism about the positioning of sex—a positioning I have not, in using top-level labels, shown—criticism regarding ethnocentrism raised by Geert Hofstede (1984) is, perhaps, valid. To go toward overcoming this, I have followed the reworking of the model by Turil Cronburg (2010) in which he finds that awareness of first-, second-, and third-person perspectives and of each one's input needs and output needs, moves through a general pattern that is basically the same as Maslow's. In accord with this, I have altered Maslow's top level from Self-Actualization to Actualization.

My own criticisms are first that there is an unacknowledged shift between needs and desires—these may overlap but are not synonymous—and second that despite any reworking to address valid criticism, the model remains analytic. By recasting needs (and desires) as values that power the real event of "mattering," a more accurate (albeit fuzzy) and comprehensive account of the interaction of the parts and the wholes can be made clear.

In *Being and Time* Martin Heidegger wrote:

> *Dasein* is a being that does not simply occur among other beings. Rather it is ontically distinguished by the fact that in its being this being is concerned about its very being. Thus it is constitutive of the being of *Dasein* to have, in its very being, a relation of being to this being. (Heidegger, 1927, 1962)

Heidegger would have been thinking of the individual person, but as identified above in discussing parts and wholes, beings, including organizations are both parts and wholes. All being is "in its very being, a relation of being to this being" and needs to be considered as such—as a relationship rather than as collections reducible to individual parts.

You and I and our dependents are human society. For human society to survive, we are also the providers of our and our dependents' needs for sustenance, shelter, safety, health, belongingness, education/learning, work, and "play." To solve the ongoing problems of meeting our needs, we create culture. I am here considering what we commonly call Western society and am therefore concerned with Western cultures. These cultures have created power structures and institutions designed to meet our needs. The primary power sources created and recognized by Western society are the state, private governance, market forces, and economic forces. The state, which welds its power through its ownership of resources, legislation, and military might, includes a spectrum ranging from dictatorship to democracy; private governance pertains to private property and to those aspects of culture over which the state allows individuals and groups to make their own decisions; market forces are those of supply and demand; and economic forces are the factors that help determine competitiveness for limited resources including fiscal and monetary policies.

Because we matter as humans within a natural environment that produces and sustains us, for survival we are both dependent on and responsible for our environment. Attention to the world's environment—much of which knows no borders—is a recently realized need for guaranteeing survival, yet responsibility for addressing this need is strongly contested at present at international forums of states such as the United Nations. Society is not bereft of cooperative abilities and indeed, in many situations, shows a high aptitude, yet compared to ecosystems, culture is woefully deficient. Some of us have over-stretched our needs (desires) at the expense of others and, now too, of our environment.

Human society is culturally out of kilter with its ultimate sustainer, the environment; as an evolutionary event its telos has become demise. This might be understandable if wealth was evenly distributed (commonwealth) throughout our world, but this we know is not the case. If we ask ourselves whether the wealth of the world is sufficient to sustain human society and to care for our environment and answer in the affirmative, then we are left wondering what is happening in such advanced economies as the United States, the United Kingdom, and Australia which is causing an increase in poverty, that is, the inability of a growing number to meet even basic physiological needs. This tragic inequality is highest in what by measures of wealth is the most advanced economy in the world, the United States (see figure 8.6).

This brings me back to consideration of the "Best Practice" Model and the "Best Fit" Model and to the part they play in the distribution of wealth and the role of meeting the needs of society which is both its maker and benefactor. How do they matter? Does power function even-handedly in each? Which evolutionary event of each model leads to a better outcome for society, for the environment, and for sustainability?

The accountants, financiers, and economists who, it is said, "run the show" these days, would, in a climate of cost minimization and high unemployment, always opt for the "Best Fit" Model as the most efficient for achieving its purpose. In situations such as the need, say, to trial a drug to combat a pandemic such as Covid-19, where demand emanates from the international organization WHO, CTCs are the best option for conducting such trials. Under these circumstances, they are the supplier most likely to have the flexibility and scientific capacity to perform when untethered by limiting contractual demands made by the pharmaceutical industry. Only in circumstances where demand for suitably qualified, capable, experienced employees outstripped the supply of such people would it be in the interest of CROs either to relax their performance limitations or offer potential employees some of the kind of incentives offered by "Best Practice" Model employers.

Any CTC that generates its own demand, as when it seeks to conduct a medium to large trial to test a hypothesis, needs to persuade an organization to fund such a trial. This can occur in cases such as BOOST (see chapter 7), or when a drug is on the market targeting one purpose, but is found to be promising in treating an unrelated condition. The latter situation is often found in oncological research, such as the use of a Cox 2 inhibitor, developed to treat osteoarthritis, to supplement the treatment of some gastrointestinal cancers. Even though cancer occurs in a large enough section of the population to warrant attention, there is a wide variety of cancers, with few responding in common to any one drug or treatment regime. It is not in the interest of pharmaceutical companies to engage in the expensive business of developing and testing drugs with the specific purpose of providing treatment for what to a global company is a small market. This is especially so since there is still big money to be made from those who can pay—"first-world" populations—for drugs for conditions such as Alzheimer's disease, or by pathologizing normal events in human growth and development and developing "treatments" for them. Nevertheless, pharmaceutical companies may be persuaded to supply a drug to a CTC for trials of secondary treatments; it is, after all, in the interest of CROs for CTCs to continue to operate given they are the training grounds for future employees of pharmaceutical companies medicines research. In some cases, where the possibility of a future new income stream is foreseen, pharmaceutical companies may even fully fund such a CTC trial without the restrictions to performance such as publishing results applying, as is often the case with the trialing of the first intended purpose drugs.

Study of the Performance Productivity Frontier in relation to the two models would suggest that the "Best Fit" Model is the better of the two. This is certainly the case when the power to put the value of profit first is considered to matter most by those who have the power to enable the flourishing of such a value. Such enablers are governments and their concomitant leaders. Given that their responsibility is to the society they are intended to represent, they would appear to have been convinced of the trickle-down effect, a theory that has never been shown to have empirical reality. It is in fact a marketing term used in support of *laissez-faire* economics which encourages governments to lower upper-echelon taxes, deregulate markets, bail out failed performers, and allow CEOs and company directors to grant themselves exorbitant remuneration packages and bonuses even when they fail. Communism, which touts itself as being for workers, appears to be no better than its most extreme rival, rampant capitalism, in serving the whole of the society for which it claims responsibility. Then again, communism cannot be said to support

Figure 8.4 The Evolution of Real Wages and GDP per Hour Worked (in the Market Sector), That Is, Labor Productivity for Australia. *Source*: Bill Mitchell. "Evolution of Real Wages" (October 29, 2010. http://bilbo.economicoutlook.net/blog/?p=21486) (Graphs: ABS June: 2010).

either model, rather it has shown itself to circumscribe freedom (and equality) and to be highly bureaucratic in practice.

Bill Mitchell, Professor in Economics at the University of Newcastle, NSW, Australia, on his blog composed a graph (Figure 8.4) using Australian Bureau of Statistics data from 2010. Similar graphs for most advanced countries over this time-line show a common trend toward an increasing gap between the wealth labor has created, represented by the upper graph (Labor Productivity) and what they were paid for producing that wealth, represented by the lower graph (Real wages). The gap represents the takings by senior management, capital owners of wealth, and the state, of that which was created by labor. Whatever way this is viewed, it reveals a lack of respect by the few for the many. When respect for people, *qua* people, and the world we inhabit as the primary concern of society is usurped by any other value, different consequences ensue. Since the early 80s, there has been a shift by some so-called advanced nations from commonwealth (commonwealth) to gross domestic product (GDP) as the primary concern of the state.

A cursory glance at a few figures should give some weight to my argument. Take the Figure 8.5 (below), for example:

The richest 10 per cent of Australians have gained almost 50 per cent of the growth in income over the past three decades as inequality has widened throughout the Western world, according to one of the world's foremost authorities. . . .

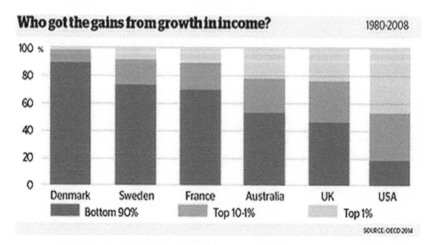

Who got the gains from growth in income? 1980-2008

Denmark · Sweden · France · Australia · UK · USA

■ Bottom 90%   ■ Top 10-1%   ■ Top 1%

SOURCE: OECD 2014

Figure 8.5   Distribution of the Gain in Six OECD Countries. *Source*: Colebatch, Tim. "Who Got the Gains from Growth in Income?" (SMH: October 10, 2013. http://www.smh .com.au/federal-politics/political-news/countrys-rich-have-lions-share-of-income-growth -20131009-2v8q2.html).

> John Martin, who has just stepped down after 13 years as the Organization for Economic Co-operation's director for employment, labor and social affairs, says new OECD figures estimate that between 1980 and 2008, 22 per cent of all growth in Australia's household income went to the richest 1 per cent. (SMH 2013, October 10)

These figures certainly put a lie to the much-touted "trickle-down effect" that, it was said, would result from the state concentrating first and foremost on deregulated economic growth. In his book, *The Price of Inequality* (2012, 2013), the Nobel economist Joseph Stiglitz describes trickle-down economics as an idea that "has a long pedigree—and has long been discredited." Far from delivering on its promise, Stiglitz points out that

> what America has been experiencing in recent years is the opposite of trickle-down economics: the riches accruing to the top have come at the expense of those down below. (Stiglitz, 2013: 8)

It is interesting to note that of the six countries compared, Denmark has the highest AAA credit rating, with Sweden not far behind. Australia holds on to its AAA by "the skin of its teeth"; France's rating is AA as is the United States; and the UK's rating comes just ahead of the last two at AA+. On the basis of figures 8.5 and 8.6, Denmark's governance is the most equitable in

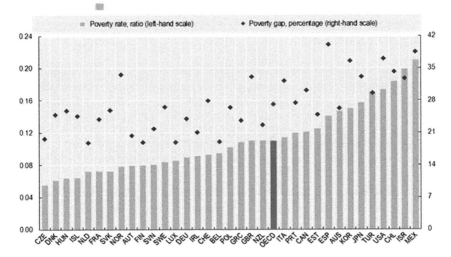

Figure 8.6   Poverty Rates and Poverty Gaps—Advanced Economies. *Source*: OECD (2010).

terms of a fair social contract without discriminating against those whose primary purpose is to make money.

Looking finally at the levels of poverty in OECD countries shown in figure 8.6 and knowing that some households below the poverty line in the United States include a person in full-time employment, respect for all people, in the sense described previously, has been outstripped by the social value of wealth creation. Speaking for the bottom 90 percent of Australian, British, and American society, many of us may have been beguiled by hype; ultimately, we've been duped.

Because this situation has evolved and is evolving in a number of OECD countries, as indicated in the last three figures, I want to spend a moment exploring the means by which this has been made possible "right before our eyes"—how a growing proportion of people in "advanced" societies are struggling to meet even their basic needs as outlined by Maslow. I have no wish to wade into political debate and therefore I have no intention of dwelling on the "whys and wherefores" of the evolving consequences. Nevertheless, because our metaphysics is where our beliefs "reside," and in order for our metaphysics to be sound it needs an interpretant. And because what Peirce called methodeutic or speculative rhetoric is of such importance to our interpretant and subsequent behavior, I will highlight its misuse, "spin," a device that can reveal or screen emerging meaning and concomitant values.

Spin is corrupted rhetoric and is ubiquitous, insidious, increasingly sophisticated, and obfuscating. According to the SOED (2002) spin, in the sense used here, is a bias or slant on information, intended to create a favorable impression when it is presented to the public (orig. *US Politics*) (L20). Lynda Mugglestone (professor of the history of English at Pembroke College, University of Oxford) adds:

> Often associated with newspapers and politicians, to use spin is to manipulate meaning, to twist truth for particular ends—usually with the aim of persuading readers or listeners that things are other than they are. As in idioms such as to put a "positive spin on something"—or a "negative spin on something"—one line of meaning is concealed, while another—at least intentionally—takes its place. Spin is language which, for whatever reason, has designs on us. (Mugglestone, 2011)

When rhetoric—a linguistic or visual device for revealing and persuading others to the teleology of ideals—is focused on winning rather than elucidation, it too easily degenerates into spin. As Mugglestone (2011) also points out, spin is often associated with newspapers and politicians and we know from the SOED that its etymology dates back to late twentieth-century US politics, but spin is just as pervasive in marketing, sales, lobbying, public relations, advertising, prospectuses, and on social media.

Unfortunately, in highly competitive societies, our general education has provided little opportunity to learn the art of rhetoric in the classical sense. We have not been taught directly how to write and speak in ways that aim to cooperatively elucidate truth. I say cooperatively because writing assumes a reader, and speaking, a listener. Without knowing much about it, we often consider rhetoric with suspicion as insincere hype. We may well have learnt in our language classes how to write eloquently; we may have studied the plays of Shakespeare without considering the *purpose* of some of his characters' great speeches, such as the differing aims of the speeches given by Brutus and Mark Antony following the assassination of *Julius Caesar* in the play of that name. We have learnt the art of debating but with the primary aim of winning and thereby, maybe unwittingly, learnt how to spin.

In science we are taught to write objectively, treating *pathos* and *ethos* as anathema to successful scientific writing, without considering that language—written, read, oral and aural—involves interpretation. Evelyn Fox Keller, physicist-turned historian and philosopher of science, addresses this issue in much of her work including *Refiguring Life: Metaphors of Twentieth Century Biology* (1995). The English philosopher Mary Midgley's *Science and Poetry* was published in 2001 and *Metaphor and Knowledge: The Challenges of Writing Science* by Ken Baake of Texas Tech University. Ken Baake, who specializes in the rhetoric of scientific literature, is at the same university as Kenneth Ketner, the Charles Sanders Peirce Interdisciplinary Professor, yet there is no indication in his book that he had ever attended the work of the Institute for Pragmatic Studies. Interestingly, in the last few pages of his book, Baake makes reference to Bryan Magee's *Confessions of a Philosopher: A Personal Journey through Western Philosophy from Plato to Popper* (1997):

> Oxford philosopher Bryan Magee delivers what might seem to be a devastating blow to my rhetorical study of science. Citing Popper's attack on positivism, Magee argues that natural science scientists produce a lot of useful information without spiraling off into irresolvable debates over what constitutes an "observation," "measurement," "light," "mass," a "number," and so forth. Nothing is more humbling than spending several years developing an argument, only to find while approaching the end that a seeming rebuttal already is in paperback at popular bookstores. (Baake, 2003: 214)

Baake goes on to say, "Magee is right, as was Popper, but this does not mean that rhetorical problems of science are irrelevant to scientists." Here I strongly agree with the second part of his statement but not the first. Magee and Popper were not so much "right" as they were transparent in their respective interpretations. Understanding this is to understand Peirce's method of inquiry. Knowing Magee's book well, I cannot agree that it undermines Baake's work—the two authors have different purposes in writing—but I can say that to my mind, Magee's rhetoric is closer to "pitch-perfect." Magee, in his 1997 book, published six years prior to Baake's, admitted to not knowing the work of the American pragmatists, but by the time his book *The Story of Philosophy* (2001) was published, he could be seen to have remedied that and was very clear in his understanding of Peirce.

Baake was emphatic about one thing: "debate over meaning is at the core of scientific work" and I am in agreement with him on this. For that matter, debate, where it is for the purpose of discovering, should be at the core of honest, open rhetoric. Let us remember, as noted previously, that for Peirce, "the *purport* of words is nothing but their *purpose*" (MS.463, 1903). For words to have meaning they must include Firstness, Secondness, and Thirdness. In like manner, inquiry must include consideration of its Firstness, Secondness, and Thirdness; its dynamical and immediate object; and its dynamical, immediate, and final interpretant—that is, application of Peirce's Pragmatic Maxim—to discover the information contained therein. Even a mathematical proof under consideration for, say, a Field prize involves a number of years of peer investigation to validate it and ascertain its value as information. When I said that Magee's rhetoric is close to "pitch-perfect" I was not implying that it is truth—as with all knowledge, it is fallible—and he did, after all, champion Popper, who had admitted regret at omitting consideration of Peirce's work. Rather I was saying that Magee's rhetoric demonstrates rigor in laying out his argument against the analytic turn, which is an underlying purpose of his 1997 *Confessions*.

I have taken the time here to discuss rhetoric—its uses and abuses—because it highlights the importance of transparency. It is not for nothing that the international organization that monitors and publicizes corporate and political corruption globally calls itself Transparency International. A lack of transparency can, as with the notion of objectivity, hide a multitude of "sins," most particularly purpose and values which are the ground of reality. Its absence, if discovered, erodes trust and cooperation and in so doing can bring out the worst in people, not least of which is disengagement, even revolt.

The "Best Practice" Model is minimally grounded in and maintained by the three values I have discussed so far: integrity, respect, and transparency. The fourth value mentioned, rigor, might be considered the spur to vigilance and, because, as Tony Jappy (2013: 73) points out, "the Peircean system precludes establishing in advance what the interpretant of a given sign may be," to evaluation. But first, the model includes two further values: commitment and cooperation. These two, together with the abovementioned braid, again with rigor as the powering of vigilance, make a stronger braid along an evolving continuum.

Returning then to my particular example of a "Best Practice" Model, a theoretical university CTC: if my three-part hypothesis of "mattering" is valid—that

1. value functions as a condition of intelligibility—"mattering" is grounded in value and is purposeful;
2. power—where power is the capacity to cause—is the enabler of force functioning as actual "mattering"; and
3. "mattering" is evolutionary realization

—and if my work, as senior administrator, is grounded in the same values identified here as essential, then, powered by appropriate, available resources, a vibrant, successful university CTC should emerge. Only continuous evaluation of the evolutionary event reveals the "mattering" of the model and allows for necessary tuning and fine-tuning of what is, after all, a human creation purpose designed for the survival—even flourishing—of humanity and, indirectly, but just as essentially, our world.

As existing, that is, *being* and *becoming*, the parts *and* the whole of that which is under consideration are "mattering." This also means that any environment is as much a whole as the parts which it supports and supports it—it is, along with its parts, organization, organized, and organizer. Hypotheses, *qua* hypotheses, as the creation of hypotheses makers, are neither true nor false. On the other hand, logic, as it is commonly understood, though truth preserving, is not truth making—unlike hypotheses, logic does not get one anywhere new. Logic does, however, when coupled with empirical validation, broadened to function semeiotically, lead from the creativity of abduction to grounded metaphysics, on to the special sciences and thence to the sciences of review, whence engagement in the practical sciences can reveal the reality of "mattering" as proposed by the hypothesis.

Let us say that if, as the senior administrator, I was to adjust the performance expectations of the university CTC in question so that they were obtainable—that is in accord with the "Best Practice" Model and mapped onto the Performance Productivity Frontier—and was able to reconcile resource requirement with resource availability, I would still need to do further work, most particularly in relation to identifying both endogenous and exogenous power, before being in a position to evaluate performance.

Given that Pfeffer's "Best Practice" Model prescribes "*[s]elf-managed teams and decentralization of decision making as the basic principles of organizational design*" (1998), this is a good place to start in revealing purposeful operation of power. Holding in mind Peirce's Pragmatic Maxim (1873 & 1905)—"Consider what effects that might conceivably have practical bearings you conceive the objects of your conception to have. Then, your conception of those effects is the whole of your conception of the object"—it is best to have a solid understanding of the concepts involved. For instance, the meaning of "team" and "decentralization" are many and vary depending on the values informing them and their purpose in so informing.

In the model, teams are comprised of people employed to achieve the organization's purpose. Pfeffer's first-mentioned practice is *employment security* (ibid.). Employment security is codified in employment contracts. The parties to employment contracts agree, on an ongoing basis, or for the term of the contract, to exchange purposeful energy and intelligence for the power of the value of remuneration, benefits, and entitlements. The fourth of the seven—*Comparatively high compensation contingent on organizational performance*—can be negotiated between employer and employees or their chosen representative/s through what in Australia are called "Enterprise Agreements." These Agreements are lodged, with the Fair Work Commission. Even given that this process of whole organization alignment may be complex and protracted in terms of ensuring clarity and comprehensiveness, enterprise agreements, once lodged, allow for appeal to external advisors or arbiters in cases of disputes concerning possible breaches and to processes of resolution. The power of a "Best Practice" Model Enterprise Agreement lies in its ability to provide a high degree of certainty to both employers and employees. For the employer it enables long-term planning; for employees it meets Maslow's first two levels of needs and can assist in the achievement of the other three levels. In his work *The Philosophy of Money* Georg Simmel notes that "the feeling of personal security that the possession of money gives is perhaps the most concentrated and pointed form and manifestation of confidence in the socio-political organization and order" (1900, 1990: 179).

Pfeffer's second practice, *Selective hiring of new personnel*, is powered, as Peirce prescribed, by imagination, concentration, and generalization. This begins with creating, building, and vertically and horizontally fitting together the roles and teams in a form that can meet the requirements of the whole as identified in the evolving purpose, mission, vision, and goals of the organization.

The raising of the flow of power through *extensive training*—Pfeffer's fifth practice—cannot be overemphasized. Pervasive training before and throughout the evolving telos of "Best Practice" Model organizations reduces the likelihood of errors and wrong turns. So too is the power of Pfeffer's seventh practice, *Extensive sharing of financial and performance information throughout the organization*, vital. In serving as a practice of transparency, it also operates as a feedback loop for the whole organization.

The power of the first part of the third practice, *self-managed teams*, requires consideration of a different kind. The only commonality of the many kinds of teams is that the people who comprise any team are all linked to a common purpose. Regardless of the kind of team, their effectiveness, that is, their ability to meet their purpose is dependent on their cognition and on the power of the second part of the practice, *decentralization of decision making as the basic principles of organizational design*. This involves the distribution of authority *throughout* the organization horizontally and vertically that is aligned with role responsibilities within an organizational matrix. In addition to a document of delegation of authority, a flowchart of authority, which includes training and which flows through performance alignment is essential.

To ensure their effectiveness, multidisciplinary and interdisciplinary teams require power *control*. Why this is so can be better understood in the light of Pfeffer's sixth principle: *reduced status distinctions and barriers*. This principle can only be achieved when there is a common understanding that every person in the organization is as important as every other person regardless of their role in the evolving purpose, mission, vision, and goals of the organization; that "mattering" of the organization is grounded not in status but in values. This will sometimes take enormous self-control by senior managers to hold back from exercising the power and authority invested in their positions and interceding arbitrarily into an organization's mutually agreed processes, priorities, and decision-making. This is not the same as engaging in change management and continuous improvement which is an inherent feature of the evolving telos of any organization. Rather it is akin to the intercession of fascism into an ostensible democracy.

Although I have not shown empirically that the "Best Practice" Model is better than the "Best Fit" Model, I hope I have shown that the latter isn't better than the former. Certainly, it is preferable to the bureaucratic way things are run in many public organizations and universities where resentment often outweighs respect, integrity is usually maintained through crisis management, and transparency by trusting no one. The "Best Fit" Model is only the "best" model where the primary purpose of an organization is to maximize the value of ordinary shares. Here the value of any reduced costs and increased productivity is being "paid" to money, the vast majority of which is in the hands of 1 percent of the population. Encouraging the more affluent of those below this 1 percent to invest their money—that is, make it do work for them—increases the number of supporters of the 1 percent.

With few exceptions, adult humans are no longer self-sufficient; we are dependent on money to exchange for goods and services to meet our needs and wants. Bereft of earning access or ability, we are dependent on government welfare, charity, or family and friends. In dire situations such as natural disasters or war, organizations such as the United Nations, the World Bank, Non-Governmental Organizations (NGOs), and, as with the present pandemic, the WHO step in to address the basic needs of those effected and affected and assist to get economies back on track. It comes to this in an economic model of society, we need to pay to live and to support our dependents and in order to pay, we need money. For the majority of the adult population, this means we need to work. As shown in the statistical tables, the few who start with or accumulate enough money to work for them—even instead of them—are a shrinking section of the population. Money flowing to these few has increased greatly over the past three decades through the increased productivity of the working population. Throughout this period, the cost of living has continued to rise but because the benefit of increased productivity has not garnered for those who labored for it, those living below the poverty line has increased.

I hope that I have shown "mattering" is grounded in values that are expressed through the enactment (or powering) of "mattering"'s purpose or telos. "Mattering" is the process of coming to be of becoming through evolutionary growth and development. It is ontological but is less about existence and more about reality. A process has consequences and ipso facto it is real. I have further attempted to show that notwithstanding that telos is a noun, through the process that is, it evolves—in a manner of speaking, it has a life of its own and grows and develops—or is overtaken. Unfortunately, our inability to recognize this, or our weakened ability to respond to it appears to be making eristic savages of us and to have set us on a trajectory of annihilation, no expense (to ourselves and our environment) spared. We, as a global society, despite any cultural differences must change our feeling, behavior, thinking and understanding, and values if we are to save our planet and ourselves.

The "mattering" of the universe, I suggest, is grounded in its telos, which I further suggest is in the value of *being* and *becoming*. As the universe continues to create itself, it is minimally necessary for its integrity, respect, and transparency to prevail. The same goes for those of us who have evolved but have yet to develop sufficiently to be fit and to fit.

Conclusion

In the vein of Peirce's strands of system, I have spun a yarn of "mattering" from its beginning by chance out of nothing to the evolving telos which is the universe. I use the ancient nautical idiom "spin a yarn" because of its multiple metaphorical meanings. To "spin a yarn" is to "tell a tale." "To tell" is "to make known by speech or writing." A "tale" is "discourse; a thing told; a story" or narrative. The wording of anything written is "text." "Text" comes from the Latin *textus* "style or texture of a work," literally "thing woven" from the past participle stem of *texere* "to weave," from the root *tek* "to weave, to fabricate, to make." Robert Binghurst, in *The Elements of Typological Style*, enlarges on this:

> An ancient metaphor: thought is a thread and the raconteur is a spinner of yarns—but the true storyteller, the poet, is a weaver. The scribes made this old and audible abstraction into a new and visible fact. After long practice, their work took on such an even, flexible texture that they called the written page a *textus*, which means cloth. (Binghurst, 2002)

Because I am human, I can create experience exogenously of what I observe, and come to believe, through language and art. If thought can be seen as a thread of consciousness, then as the storyteller I am spinning long threads together into yarn, which is spun thread, then taking these strands of yarn and weaving them into a narrative. A true storyteller, however, is not necessarily a poet. Take my poem, "Not Guilty":

I'd be deluding you
if I said I was lying.

Don't be deceived!

The fact of the matter is
I don't know what
bit me on the arse.

If the truth be known,
it should be locked up
and I should be set free.

Insomuch as it is a mental creation, this plea of innocence is a hypothesis. Insomuch as it is a hypothesis, it is neither true nor false. Insomuch as it is poetic, the purpose for its creation is the created image itself. In this case my purpose was the creation of an image of the proposition: "I'm not the cause of the effects of my actions/decisions/choices." My interest was in exploring the ways humans deny existential responsibility. In pointing out the difference between the interest of poets in creating hypotheses and the interest of mathematician engaged in the same sort of creative action, Peirce wrote,

> the poet is interested in his images solely on account of their own beauty or interest as images, while the mathematician is interested in his hypotheses solely on account of the ways in which necessary inferences can be drawn from them. (PM. 91-2, 1903).

Unlike the proposition of a poem, a mathematical hypothesis is a theory in need of verification. Peirce's philosophy is a method, a system—an architectonic—for discovering, evaluating, and interpreting reality.

As with Peirce's architectonic, in spinning the threads that make the yarn for weaving the fabric of my narrative, I have begun mathematically, that is, I have observed "rhythms of the universe," and framed my hypothesis of "mattering" accordingly. My narrative, although supported by a growing body of strong empirical evidence, is nonetheless a philosophical account of "mattering." It is a metaphysical tale and is so because while I can support its validity, I cannot attest to its truth. To know anything, we need data; understanding requires information. The information necessary to understand what I mean by the reality of "mattering" requires first an understanding of the philosophical work of Peirce. I discussed truth as being *en futuro*, then confronted the question of what we believe, concluding that, despite protestations to the contrary, we all believe. I then enlarged on the subject of information. Following that, with some trepidation, I approached cosmogony and the West's idea of the creation of the universe. I discussed what are sometimes called the religions of the Book—Judaism, Christianity, and Islam—covering the period from translation of Aristotle into Latin by Islamic falsafa scholars, through to the end of the scholastic period and the rise of nominalism. Meanwhile, the universe continues to create itself, and it is this that the cosmologists work to discover, understand, and explain. They have yet to find an adequate solution to the enigma of how the universe holds together.

Peirce's metaphysical term "agapasm," as the mediator between tychism (chance) and synechism (continuity), is supported by Christian revelation as recorded in the New Testament. Referring to the Second Letter of Paul to the Corinthians, from chapter 13, verse 8-13: perhaps it is as Paul wrote.

> Love will never come to an end. Are there prophets? their work will be over. Are there tongues of ecstasy? they will cease. Is there knowledge? it will vanish away; for our knowledge and our prophesy alike are partial and the partial vanishes when wholeness comes. When I was a child, my speech, my outlook and my thoughts were all childish. When I grew up, I had finished with childish things. Now we see only puzzling reflections in a mirror, but then we shall see face to face. My knowledge now is partial; then it will be whole, like God's knowledge of me. In a word, there are three things that last forever: faith, hope and love; but the greatest of them all is love.

All scientists, regardless of their religious beliefs or leanings could take some of this "on board": all could agree that "our knowledge and our prophesy [predictions] alike are partial" and even go so far as to agree, "the partial vanishes when wholeness comes." They may concede to the first part of verse 12: "[n]ow we see only puzzling reflections in a mirror," but, as scientists, might view the rest with skepticism. Word swapping is a short-term solution. Peirce uses "agape"; the King James edition, "charity"; in the edition of the New Testament cited above the translation is "love." It is also unfortunate that namby-pamby images are often conjured up of love (luv). Perhaps when physicists discover the mediator of quantum theory with the atoms of matter, we will have a fitting metaphor.

As a continuation along the evolutionary continuum of mattering, humans are determined, but to the degree that we have developed our ability to think, we have regained some of the freedom foregone as the means of meeting the telos of "mattering." As we learn to weld this which we call free will, we have gained responsibility and thereby the ability to both adorn and undermine the primary purpose of "mattering." One of the earliest words in a child's vocabulary is "no," a word expressing the great power of free will and it is one that is an expression of what I mean by responsibility. Later small children learn another practice of power which they take great delight in pressing the button—whatever button makes something happen. One of the defining characteristics of life is that it reacts, unlike all "mattering" before it which has surrendered in order to matter. Humans, through their ability to think have developed the ability not only to react but also to respond; to the extent that we have free will, we alone, in this world at least, are responsible: we can choose. This ability humans alone have developed—to extend beyond reaction and response—is our aesthetic and moral ability.

Our moral ability is not simply a matter of choosing between right or wrong, good or bad. Peirce, as I have detailed, eventually chose Normative Ethics and, as indicated in his Pragmatic Maxim, a kind of consequentialism. Yet his consequentialism is not that of Bentham's utilitarianism, bereft as it is of Firstness. In the absence of Firstness, there is nothing from which to prescind when inquiring or choosing and we are left with only those unsatisfactory methods of fixing belief described in "Fixation of Belief" (Peirce, 1877): tenacity, authority, and a priori. Peirce emphasized that his method is fallible—there is no certainty—but that it is the surest method for getting to the truth of "mattering."

All choosing has purpose and is grounded in values. I have spoken of the values of commitment as expressed in our primary purpose and of integrity, transparency, and respect. These same values are all inherent to the evolutionary telos of the universe. They are necessary for the production of information but are not sufficient. I suggest that the addition of the values of imagination, mindfulness, and cooperation are also necessary to meet the requirements for the creation of information which is "mattering."

I would also like to see some genuine doubt generated in overly confident self-righteous minds. This, though, is wishful thinking on my part. As Jackie Polzin says in her novel *Brood* (2011, p. 88),

> I never feel smaller than when I am filled with doubt, such a small, small feeling, it's a marvel it can fill anything at all. Filled with doubt I shrink until I can hardly move, can do nothing but wait and see what happens.

The world, especially the United States, has experienced the dreadful consequences of a charlatan being in a position of the greatest bestowed authority. Although fully fledged skepticism is not to be recommended neither is violence a means of imposing will. I would venture here that rekindling curiosity—a primary driving force of our early learning—would not go astray.

The potential for a tale streams forth spontaneously from nothingness; it is created, creation, and creativity. Its purpose is to be and its value is in it becoming so. This is its Firstness. A real narrative is not only factual: the fact of "mattering" is its Secondness. A true story can only be indicated in the long run: the truth of "mattering" is its Thirdness and is happening as I write. If the truth of "mattering" be known, it must be experienced. Through trust, that is, suspension of disbelief—in this case, the suspension of disbelief in the validity of subjectivity in a scientific study—and via the method of endophysics as discussed by Kirsty Kitto (2006, 2007) we can engage semeiotically with the subject of our study, the Cosmos, in its Firstness, Secondness, and Thirdness; we can experience the evolutionary telos in which we are partaking.

As bound up in the universe, we are also determined by and determining its self-creation. Yet, because as humans we can choose, our discourse also includes the possibility of denial, subterfuge, and lies; to the great detriment of our world and ourselves, we flout limit—we disengage from "consentment"—and we think we can get away with it. I have exemplified denial in my poem *Not Guilty*. This can be seen in the outrageous cries of climate-change deniers and of those who would concede to climate change but still deny all responsibility for its acceleration. These and many more also deny responsibility for ravaging our environment, because, quite simply, it is not in their interest to do so. Subterfuge is exemplified by the behavior of the parties' intent on bringing to market the industry-developed drugs called Cox 2 inhibitors before they had been fully tested, as discussed in chapter 8. As to lies, look no further than the so-called "trickle-down" which continues to be believed by vast numbers of the 99 percent of humanity who are its victims. The lie is not in the original exploration of it as hypothesis but in its projection as evidenced, by those who benefit from its perpetuation.

Free will means we can change this untenable trajectory. Peirce wrote:

> We know only too well that all things are not just as we would like to have them; and the only way of improving the situation is to do something about it: mere dreaming will not answer the purpose. This is what is meant by saying that things are real. If the world were a dream, we could just dream otherwise. But the real is that which is as it is whatever you or I may think about it. (PM. 132, 1906)

The real is what matters. It does not matter because you or I say it matters but because it is real. You and I matter; we matter as much as any sun or moon but no more than any grain of pepper or drop of water. So too do our creations— those things we do—matter: growing, preparing, and eating food; designing, making, and wearing clothes; building shelter, workplaces, schools, hospitals, and communities; caring for one another and our environment in sickness and in health; crafting, engineering, and communicating; and creating and engaging with our plethora of arts. And we play. Nonetheless, we also engage in denial, subterfuge, lying, violence; building barriers; and making, selling, and using weapons. All these and more—all that is—matters. All have potential and are purposeful, grounded in value, actualized through their creation, and have trajectories of possibility—all is real: it is "mattering."

So much of our part in creation is truly wonderful: hot coffee, a red frock, a veranda on which to sit in a red frock, drinking hot coffee with a friend, the thoughts we bandy about, the memories we recall, the ideas we explore, the music from the radio of the saxophone, and the guitar of Col Loughnan and Steve Murphy playing the latter's composition The Coogee Shuffle.

All this matters and is "mattering," but so too is gross misuse of power: greed, lashing out, the destruction, denuding, poaching, carnage, annihilation, scorched earth, genocide. These are not glorious. For some of us, our freedom to choose and values are haywire.

We are, therefore, we feel—we experience, we learn, we think, we value, we choose, and we do. We learn to think in the same way we learn to read: by doing it. The more we do it, the better we become at it. We need to do more—much more—of both: broadly then deeply so we can move beyond knowledge to understanding of "mattering." We also need to highly value teaching and learning for meaning and always with a view to understanding the whole and the parts and how they fit. In this, it is not enough to master the "Three Rs"; we need the arts, especially the one most accessible to everyone like music, in addition to philosophy, history, the many and varied sciences, and physical engagement.

Regaining the livelihood of all: curbing any penchant we have for violence and violation, eroding trust, extremism, and dogma; securing the livelihood of life; and reclaiming our environment—our home and our planet Earth—are the things we need to do to change the trajectory of our shared responsibility of "mattering." The universe can do it and, even granted that it has been practicing forever, we too can do it: we can make it continue to matter. I don't know what we do to recover from the narcissism, self-centeredness, selfishness, self-righteousness, greed, hate, and nihilism which plague so many of us and which lead to such destruction. For my part, I'm presently exploring power, equality, intersectionalities of society and culture, and friendship.

My pragmatic best toward taking responsibility for reality has been to commit my theory of "mattering" to paper. The whole/parts creating, creation, creator is universal mattering. It is reality and it matters because it is "mattering."

Amen.

Appendix A

Peirce's Architectonic Classes

MATHEMATICS

Publication of two books in 2010, both edited by Mathew E. Moore, *Philosophy of Mathematics: Selected Writings Charles S. Peirce* and *New Essays on Peirce's Mathematical Philosophy* have made possible the task of reviewing, with some confidence. (This is not to detract from the four volumes of *The New Elements of Mathematics* by Charles S. Peirce edited by Corolyn Eisele [1976] which I was never able to buy or borrow.) Viewing mathematics as the first heuretic science, (see Figure A1.1) Peirce states that mathematics

> does not undertake to ascertain any matter of fact whatever, but merely posits hypotheses and traces out their consequences. It is observational, in so far as it makes constructions in the imagination according to abstract precepts and then observes these imaginary objects, finding in them relations of parts not specified in the precept of construction. (CP1.240, 1896)

A number of the papers in *New Essays on Peirce's Mathematical Philosophy* (2010b) detail Peirce's thesis of mathematics as observational. Let me here refer at length to the contribution by Daniel Campos (pp. 123-145).

Peirce, in his 1902 paper *The Essence of Mathematics*, restated that according to its method, mathematics is "the science which draws necessary conclusions" (CP4.228, c.1902) then gave a second, complimentary definition which, according to its aim and subject matter, is "the study of what is true of hypothetical states of things" (CP4.233, c.1902). In considering these complimentary definitions Campos shows that two kinds of mathematical hypothesis-making can be identified: what he called *framing* and *experimental* hypotheses. Peirce distinguished between the two kinds of necessary reasoning—corollarial and theorematic—which originated in his study of the

SCIENCE

> **HEURETIC SCIENCES** *[branch]*
>> **Mathematics** *[class]*
>>> Finite collections *[order]*
>>>> Pure deductive logic *[suborder]*
>>>> General theory of finite collections *[suborder]*
>>> Infinite collections *[order]*
>>>> Arithmetic *[suborder]*
>>>> Calculus *[suborder]*
>>> Continua *[order]*
>> **Philosophy** *[class]*
>>> Phenomenology *[order]*
>>> Normative sciences *[order]*
>>>> Esthetics *[suborder]*
>>>>> Physiology *[division]*
>>>>> Classification *[division]*
>>>>> Methodology *[division]*
>>>> Ethics *[suborder]*
>>>>> Physiology *[division]*
>>>>> Classification *[division]*
>>>>> Methodology *[division]*
>>>> Logic / Semeiotic *[suborder]*
>>>>> Speculative grammar *[division]*
>>>>> Critic *[division]*
>>>>> Methodeutic *[division]*
>>> Metaphysics *[order]*
>>>> Ontology *[suborder]*
>>>> Religious metaphysics *[suborder]*
>>>> Physical metaphysics *[suborder]*
>> **Special sciences** *[class]*
>>> Physics *[subclass]*   Psychics *[subclass]*
>>>> Nomological sciences *[order]*
>>>> Classificatory sciences *[order]*
>>>> Explanatory sciences *[order]*
> **SCIENCES OF REVIEW** *[branch]*
> **PRACTICAL SCIENCES** *[branch]*

Figure A1.1   Peirce's Perennial Classification of the Sciences. *Source*: Adaptation of Kent (1987: 134–135).

structure of proofs in Euclid. A corollarial deduction does not involve any ideas other than those already stated in a premise, that is, there is no need to imagine any formal relations not stated in the original premise. Theorematic proof "differs from a corollarial proof from a Methodeutic point of view, in as much as it requires the intervention of an idea not at all forced upon us by the terms of the thesis" (N4.8, 1901). He argued that "theorematic reasoning characterizes mathematics as an activity" (CP4.233, 1902). The first step of this activity is to "demonstrate that" the second step to *imagine* an individual diagram that *embodies* all the general characteristics assumed in the hypothesis; the third step is to experiment upon the diagram by *imagining*—and actually drawing, if necessary to aid the imagination—modifications that might help us to *show* that the hypothesized relations do obtain. This "experimentation" consists of the imaginative and judicious modification of the original diagram so as to produce a related diagram that might literally show or *monstrate* the hypothesized relations among elements of the original diagram. The introduction of an experimental hypothesis is the theorematic step. The consideration of possible experimental diagrams is constrained by the assumptions of the mathematical system within which the mathematician is working. In Euclid's geometry these assumptions are given by postulates and common notions or axioms. In general, a postulate is "the affirmation of a possibility" while an axiom "is the denial of a possibility" (N4.8, 1901). They constrain, then, the possibilities for creating diagrams of relations within the given mathematical system, including diagrams that represent experimental hypotheses in the course of a demonstration. Theorematic reasoning is not confined to geometry—but takes many other forms in other mathematical problem contexts.

As previously suggested, the design and construction of the Sydney Opera House serve as an example of the workings of Peirce's Classification of the Sciences and in this case, of mathematics. Not only were the drawings rough, but Utzon's submission was short on detail. It became very obvious to the appointed engineer Ove Arup that much of the design was impossible to construct. The shells, for example, were originally a series of parabolas, which couldn't be built with the technology of the time.

Moving up a level to the Sciences of Review (motivated by discovery for the sake of applying knowledge) Utzon and Arup found no precedent to act as a guide; they had to return to first principles for a solution. Paul Bentley quotes Yuzo Mikami, an architect on the design teams of Utzon and Arup who tells the story of finding the solution.

One summer day in 1961 Utzon . . . began dismantling the perspex model, sadly thinking that it would have no use if he could not find a solution for it to be constructed in a rational way.

An idea flashed in his head like a lightning . . . [W]hy couldn't they be cut out from a common surface? In order to do that the curvature must be the same in all directions. What is a geometrical body with a constant curvature in all directions? A sphere!

He rushed home and taking a child's rubber beach ball, put it into the bath-tub full of water. . . . After many trials he realized that the variety of shapes and sizes available were almost limitless. Big and small, flat and upright. He could now compose the whole shell by the pieces of spherical triangles cut out from just one single sphere. He had found the solution. (Bentley, 2001, 2013)

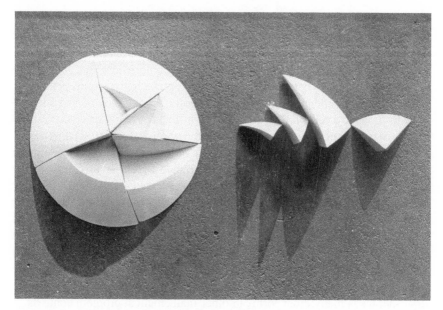

Figure A1.2   Model of Jørn Utzon's Spherical Solution. *Source*: © Mitchell Library, State
Library of New South Wales and Courtesy Jørn Utzon.

PHILOSOPHY

In considering philosophy as a science rather than an art, Peirce was treating
it as a method for discovering truth—as a process—not as a *Weltanschauung* which is often a product. For him, philosophy and its components are
considered

> derivative of formal sciences because they do not study the form of their own
> constructions but study the form of things already constructed, so to speak. One
> might label them "reconstructive" formal sciences rather than "constructive"
> ones like mathematics. (Liszka, 1996: 2)

In contradistinction to the special sciences which discover new phenomena,
however, Peirce stressed that philosophy "limits itself to so much of truth as
can be inferred from common experience" (CP1.184, 1903). By "experience"
he explained, "I mean any conscious effect contributing to a habit, satisfactory
and self-controlled yet destructible by no exercise of intellectual vigor but
only by deadening one's powers" (MS.843, nd). John E. Smith points out that

> the key to Peirce's conception of experience is found in his rejection of
> privileged starting points for thought and his emphasis on the *insistence* of

experience over against the human will, including elements of surprise which confound human expectations. (Smith, 1978: 91)

Two papers by Peirce that appeared in the *Journal of Speculative Philosophy* (2: 1868), "Questions Concerning Certain Faculties Claimed for Man" (pp. 103-114) and "Some Consequences of Four Incapacities" (pp. 140-157), are of particular relevance to his pragmaticist philosophy. So too are his two papers published a decade later in *Popular Science Monthly*, 12: "Fixation of Belief" (1877: 1–15) and "How to Make Our Ideas Clear" (1878: 286–302). In arguing his position, he took particular issue with Descartes and with the British empiricists John Locke, George Berkeley, David Hume, and John Stuart Mill. He rejected the foundationalist doctrine of Descartes, arguing that knowledge could not be regarded statically as a body of propositions but, as with the notion of scientific progress which had become commonplace in the nineteenth century, dynamically as a process of inquiry.

[Science] is not standing upon the bedrock of fact. It is walking upon a bog and can only say, this ground seems to hold for the present. Here I will stay till it begins to give way. (RLT.176-7, 1898)

So too, he argued, was this the case for epistemology. In arguing against Descartes's program for finding certainty, Peirce also rejected Hume's skepticism. Peter Skagestad summarizes:

If we have no independent vantage point from which to *justify* our body of knowledge as a whole, neither do we have an independent vantage point from which to *reject* our knowledge as a whole. The only vantage point available is the one provided by the knowledge we in fact have. Within the framework of this knowledge we may criticize and reject each one of our beliefs individually, but we cannot step outside this framework and reject all our beliefs collectively. Total skepticism is as impossible as absolute certainty. (Skagestad, 1981: 19)

His issue with Locke was the latter's notion of *tabula rasa*. Peirce agreed with his definition of experience but found it wanting in two respects. In assuming *tabula rasa*, it discounted past experience and denied expectation and thus did not account for surprise. Observers are neither passive nor "pure," likewise for observables. This is so even with objects being observed. In speaking of matter as "effete mind" (Peirce, 1891), Peirce was acknowledging that even so-called inert matter is active, albeit languidly so. However, as John E. Smith points out, Peirce "insisted that the kind of thing to which the word 'experience' is more particularly applied is an 'event' which has a temporal stretch or duration" (1978: 93).

In this regard, Peirce also strenuously opposed the doctrine of intuition at the base of John Stuart Mill's empiricism. Having shown that "there is no intuition or cognition not determined by previous cognitions," he argued, "it follows that the striking in of a new experience is never an instantaneous affair, but is an event occupying time and coming to pass by a continuous process" (EP1.39, 1868).

And while he admired Berkley, he argued against his view of reality as actual perception and his neglect of real possibility.

Phenomenology

Phenomenology (phaneroscopy, as he first named it) the first order within the derivative or reconstructive formal heuristic class, philosophy, should not be confused with the continental version as developed by Edmund Husserl, though there are similarities. According to Peirce:

> This science of Phenomenology is in my view the most primal of all the positive sciences. That is, it is not based, as to its principles, upon any other positive science. By a positive science I mean an inquiry which seeks for positive knowledge; that is, for such knowledge as may conveniently be expressed in a categorical proposition. (PAP. 120-121, 1903)

Herbert Spiegelberg (1956: 168) cites a letter to James dated October 3, 1904, in which Peirce argued that what he was calling phenomenology was, as he saw it, what William James was propounding under the new title of radical empiricism.

> As I understand you, then, the proposition you are arguing is a proposition in what I called phenomenology, that is just the analysis of what kind of constituents there are in our thoughts and lives (whether these be valid or invalid being quite aside from the question). . . . Perhaps the most important aspect of the series of papers of which the one you sent me is the first will prove to be that phenomenology is one science and psychology a very different one. . . . Phenomenology has no right to appeal to logic, except to-deductive logic. On the contrary, logic must be founded on phenomenology. Psychology, you may say, observes the same facts as phenomenology does. No. It does not observe the same facts. It looks upon the same world and the same world that the astronomer looks at but what it observes in that world is different. Psychology of all sciences stands most in need of the discoveries of the logician, which he makes by the aid of the phenomenologist. (CP8.295-297, 1904)

Peirce's phenomenology—his doctrine of categories—"ascertains and studies the kinds of elements universally present in the phenomenon; meaning by the *phenomenon*, whatever is present at any time to the mind in any way" (CP1.186, 1903). It requires keen skills of observation. He initially developed this crucial aspect of his philosophy in response to the categories of Kant. Likewise, he considered those of Aristotle whose work he had also studied in great depth. William L. Rosensohn summarizes Peirce's lead-up to the development of his own doctrine of categories as follows:

> According to Peirce, Kant fails to undertake any pre-logical analysis prior to accepting a table of logical judgements from which his set of categories could be deduced. Thus "he affords no warranty for the correctness of the original table" [W1.351, 1866]. Aristotle commences with "a half-grammatical, half-logical analysis" [L75, 1902] of composite forms of speech, deriving a list of ten types of simple linguistic expressions predicable of a subject. These he called categories, or predicaments. Peirce, however, makes his "one contribution to philosophy" [L67, 1905] neither with logic nor language but "experience" itself. His categories, he contends, rest on no "previous philosophizing at all" [L75, 1902]. By going "back to experience, in the sense of whatever we find to have been forced upon our minds" [L75, 1902] he introduced a new method: phenomenological analysis, or the analysis of the phenomenon. (Rosensohn, 1974: 37)

The principles of mathematics inform his phenomenology. As stated earlier, mathematics is "the study of the substance of hypotheses, or mental creations, with the view to the drawing of necessary conclusions" (Moore 2010b: 4). Peirce's hypothesis is that the categories, which he called First, Second, and Third, constitute the passage from Being to Substance. While, the first four pages of his paper "The Logic of Mathematics: An Attempt to Develop My Categories from Within" (MS.900, c. 1896; CP1.417-520) are missing, enough has survived to judge that his mental creation is sound. Lest seeing "logic" in the title, one is inclined to put the cart before the horse, Peirce brings to our attention that

> mathematics performs its reasonings by a *logica utens* which it develops for itself and has no need of any appeal to a *logica docens*; for no disputes about reasoning arise in mathematics which need to be submitted to the principles of the philosophy of thought for decision. (PAP.157-8, 1903)

Peirce tells us that the purpose of phenomenology is to describe what is before the mind and to show that the description is correct. He further tells us that what we have to do, as students of phenomenology, is simply to open our mental eyes and look well at the phenomenon and say what are the characteristics that are never wanting in it, whether that phenomenon be something that outward experience forces upon our attention, or whether it be the wildest of dreams, or whether it be the most abstract and general of the conclusions of science. (PAP. 152, 1903)

In a letter to the Italian pragmatist Mario Calderoni, speaking of the phenomenon, or what he came to name the phaneron (another of his neologism), he wrote:

I use the word phaneron to mean all that is present to the mind in any sense or in any way whatsoever, regardless of whether it be fact or figment. I examine the phaneron and I endeavor to sort out its elements according to the complexity of their structure. I thus reach my three categories. (CP8.213, c.1905)

Joseph Esposito (1980) gives a detailed study of Peirce's categories, but as an initial guide to recognizing these, let me first give a brush-stroke sketch of the genuine form of the three, which I will discuss in more detail as I proceed:

Firstness is potential, possibility, may be; it is a qualifier, feeling, sensation, spontaneity; in logic it is terms; in Peirce's semeiotic it is icon; in his metaphysical doctrine it is *tychism* (Greek for chance). The conventions of those of our languages which insist on the copular enable us to speak of Firstness, but while logically it is positively characterized, negatively it is devoid of existence. As Peirce says of Firstness, "*It is without reference to anything else* within it or without it, regardless of all force and of all reason" (CP2.85, c.1902).

The idea of First is predominant in the ideas of freshness, life, freedom. The free is that which has not another behind it, determining its actions; but so far as the idea of the negation of another enters, the idea of another enters; and such negative idea must be put in the background, or else we cannot say that the Firstness is predominant. Freedom can only manifest itself in unlimited and uncontrolled variety and multiplicity; and thus the first becomes predominant in the ideas of measureless variety and multiplicity. (CP1.302, c.1894)

Secondness is brute force; it is habit, resistance, dependence, contingency; it is experienced as otherness; in logic it is propositions; in semeiotic, indices; in Peirce's cosmology it is *synechism* (Greek for continuity).

Secondness consists in one thing acting upon another—brute action. I say brute, because so far as the idea of any law or reason comes in, Thirdness comes in.

When a stone falls to the ground, the law of gravitation does not act to make it fall. The law of gravitation is the judge upon the bench who may pronounce the law till doomsday, but unless the strong arm of the law, the brutal sheriff, gives effect to the law, it amounts to nothing. True, the judge can create a sheriff if need be; but he must have one. The stone's actually falling is purely the affair of the stone and the earth at the time. This is a case of reaction. So is existence which is the mode of being of that which reacts with other things. (CP8.330, 1904)

Thirdness is reasonableness; it is purpose, order, mediation, law, and adaption; in logic it is inferences; in semeiotic, symbols; in metaphysics, *agapasm*.

A third has a mode of being which consists in the Secondness that it determines, the mode of being of a law, or concept. Do not confound this with the ideal being of a quality in itself. A quality is something capable of being completely embodied. A law never can be embodied in its character as a law except by determining a habit. A quality is how something may or might have been. A law is how an endless future must continue to be. (CP1.536, 1903)

Because Secondness is an essential part of Thirdness and Firstness is an essential part of both Secondness and Thirdness, there is a distinction between the genuine and the degenerate forms of the categories. Vincent Potter lists the following combinations by example (there are others):

1. Firstness of Firstness—quality in itself, or possibility (*Primity*)
2. Firstness of Secondness—existence or actuality (*Secundity*)
3. Firstness of Thirdness—mentality

Taking Secondness:

1. Secondness of Secondness—reaction
2. Secondness of Thirdness—law as actual compulsion

Finally, Thirdness:

1. Thirdness of Thirdness—generality, lawfulness, reasonableness. (Potter, 1997: 17)

As Felicia Kruse points out, however, in her paper "Genuineness and Degeneracy in Peirce's Categories":

> The categories of Firstness, Secondness and Thirdness express relations, not absolute entities. Because genuine Secondness and Thirdness are relations, they are dependent for their genuineness upon the terms that maintain the relations. . . . The distinction between genuineness and degeneracy in the Peircean categories is perhaps best understood as a distinction along a continuum. On one end of the continuum we have the most genuine cases and on the other we have the most degenerate—those which are "merely a way of looking at things." In between, we have the realm of the relatively genuine and the relatively degenerate, where genuineness and degeneracy depend not only upon the nature of the terms of the relations themselves, but also upon the position of the relations with respect to other relations. (Kruse, 1991: 292–3)

In order to describe phenomenon, we need first to engage in mental separation of "thisness" from the "thusness" of the phaneron. According to Peirce, when observing the phaneron, there are three ways of attending to aspects or parts of the whole. These three modes of mental separation are prescision, distinction, and dissociation, with the first being the method of abstraction.

Let me exemplify this through exploration of the following incident related to the BBC science program *The Human Mind* (Spiers & Gimple, 2003) and given as an example of the operation of the psychological phenomena, intuition: in 2001 Andy Kirk, a UK fire chief, was called out to a big factory fire. Judging it safe enough for his experienced team to manage, Kirk sent them in to extinguish the fire. Within a short period, the team appeared to have it under control. Kirk then had what he later described as a "strange feeling." Sometime later he was able to piece together the mediation process between his sudden "strange feeling" and his response. His cortex had picked up changes in his present situation; he remembered every fire he had ever experienced and identified three differences between them and the scene before him: the smoke was orange, air appeared to be moving into the building rather than out as was usual, and whereas fires usually crackle, there was no sound.

Let us say, for the sake of the exercise, that an artist, seeing the surprising event of orange smoke, wants to paint it. For a painter, the meaning of orange smoke would differ from that of a firefighter. A trained artist looks at a scene in terms of the elements and principles of design. The elements—line, shape, direction (horizontal, vertical, and oblique), size, texture, and color (shade, hue, chiaroscuro, and intensity)—are the things that form design. The principals, which govern the relationships of the elements, are balance, proximity, alignment, repetition, contrast, and space. Concentrating here on the element of color and being fully conversant with the color wheel, the artist can distinguish colors and knows that gray, which could be seen because of its being embodied in smoke, is tertiary with cyan predominating. Seeing it as orange, the artist would prescind magenta and yellow and dissociate cyan.

Nevertheless, observing smoke oranging, the artist may experience a sense of, say, foreboding and create an artwork that, juxtapositing the elements and principles of design imaginatively, paints the experience of foreboding. The aesthetic "beauty" of the painting is "measured" by the degree to which it embodies feelingfulness creatively.

To a fire chief, such observation is moot; their purpose in observing such a scene and therefore how they view it is significantly different from that of an artist. Although for both, orange smoke is a surprising fact and generates a feeling—maybe one of foreboding—the purpose of observing what is going on around them differs. The feeling of, say, foreboding the fire chief Andy Kirk may have experienced may well have arisen from a subconscious awareness that "things" didn't seem to fit with what his crew had reported. As their chief, his purpose is to ensure the firefighting team performs effectively, efficiently, and ethically; that they meet health and safety standards to care for their own well-being; and that they prioritize saving life over property in the performance of their job. More than this, though, which is more in keeping with the role of a supervisor, as chief, Andy's role would have included taking a wider view of events—the bigger picture—and responding to it appropriately in accordance with his purpose.

By way of explanation, let me return to the incident in more detail: taking the second surprising fact the fire chief picked up on that air appeared to be moving into the building rather than out as would be expected. Knowing that air is colorless, I am assuming that the presence of smoke made this observation possible. Furthermore, because he makes no reference to wind, I presume it was not a factor in causality. Having identified smoke moving into the building, names have been named and thus the act of distinction completed. Dissociation serves only indirectly in that in the absence of wind, one would not usually associate smoke moving toward the fire that had produced it. While distinction is useful for describing this situation and dissociation (indirectly) indicates a surprising fact observed, the mode of mental analysis that can be used to interpret the available data, that is, to construct an explanation (abduct) is prescision. Thus, though reference to movement cannot be prescinded from air, air can be prescinded from movement, and movement thereby considered in relationship to air. The constituent of air that fuels fire is oxygen; as fire burns oxygen, air pressure is reduced and more air moves in. If for some reason oxygen cannot be replaced at the same rate it is depleted, then pressure increases to the extent that in the presence of smoke, the movement of air into the building stands out from all other everywhere-and-at-all-time movement of air. The fire chief was, in fact, observing the operation of localized air pressure.

His third observation—that the crackling sound normally associated with large fires was absent—is less straightforward to interpret. Peirce tells us "prescission consists in . . . supposing one of the two constituents of the idea, termed the prescinded constituent to be realized in a subject while the other, termed the abstracted constituent is supposed to be absent" (MS.284, 1905). The abstracted constituent that was absent was sound and it was this that finally alerted the fire chief to the fact that it was not fire that he was witnessing but a lack thereof. Kirk's observations had picked up on the signs of an imminent back draft. He pulled his team out and moments later the factory exploded.

Peirce's three categories crystallize observation in every class—in the superordinate mathematics and the subordinate, normative sciences, metaphysics, the special sciences, the sciences of review, and the practical sciences—by serving as the means for making experience intelligible. They are universal and elementary and are purely formal. "They are elementary, because they are the constituents of all experience; universal, because they are *necessary* for *any* understanding" (Rosensohn, 1974: 45).

Phenomena experienced are not limited, as the classical empiricists would have it, to those detectable by the senses. When describing what he came to call the *phaneron*, as "whatever is present at any time to the mind in any way"

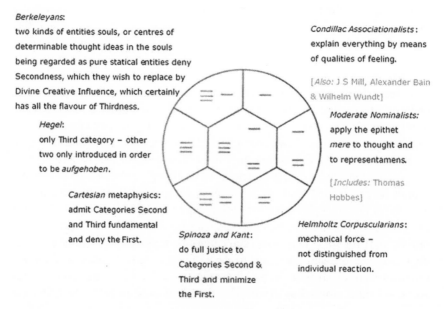

Berkeleyans:
two kinds of entities souls, or centres of determinable thought ideas in the souls being regarded as pure statical entities deny Secondness, which they wish to replace by Divine Creative Influence, which certainly has all the flavour of Thirdness.

Hegel:
only Third category – other two only introduced in order to be *aufgehoben.*

Cartesian metaphysics:
admit Categories Second and Third fundamental and deny the First.

Spinoza and Kant:
do full justice to Categories Second & Third and minimize the First.

Condillac Associationalists:
explain everything by means of qualities of feeling.

[*Also:* J S Mill, Alexander Bain & Wilhelm Wundt]

Moderate Nominalists:
apply the epithet *mere* to thought and to representamens.

[*Includes:* Thomas Hobbes]

Helmholtz Corpuscularians:
mechanical force – not distinguished from individual reaction.

Figure A1.3 Peirce's Diagram of the Seven Systems of Metaphysics, through Reference to Firstness, Secondness, and Thirdness, Summing Up the Shortcomings of Six of the Systems. *Source:* Adaptation of CP5.78, 1903.

(CP1.186, 1903), he later added "quite regardless of whether it corresponds to any real thing or not" (MS.1334, 1905). By way of example, in what Peirce identified as the seven systems of metaphysics, through reference to First-ness, Secondness, and Thirdness, he identifies the shortcomings of six of the systems (CP5.78, 1903).

Prescision is multiple and entails identification of those aspects of the phaneron that will provide sufficient means to inquirers enabling them to discover probable explanations that matter to the motives, ideals, and purposes of their lines of inquiry. As a practical scientist, the fire chief is motivated to inquire for the sake of doing. But doing what? This is where the next division of Peirce's classification, the normative sciences come in, addressing the regulatory methods for finding answers.

Normative Sciences

Peirce did not settle on the normative science as the mid-class of philosophy until c. 1903. He came to realize that "though these sciences do study what ought to be, that is, ideals, they are the very most purely theoretical of purely theoretical sciences" (CP1.281, 1903). For him,

> The three normative sciences are logic or semeiotic, ethics and esthetics, [being the three doctrines that can be distinguished on the normative curve]; Logic in regard to representations of truth, Ethics in regard to efforts of will, and Esthetics in objects considered simply in their presentation. (CP5.36, 1903)

To be in a position to make evaluation is to make use of some criterion. Normative science, he contends, "investigates the universal and necessary laws of the relation of Phenomena to Ends" (CP5.121, 1903). Vincent Potter points out, "Peirce looks upon normative science as positive science" (1997: 25).

Viewing the normative sciences as natural kinds, Peirce may appear to commit the naturalistic fallacy, but this is not so. Concerning Peirce's theory of the formation of habits, imperatives are categorical; through the process of translating this theory into practice, these imperatives become hypothetical. Exploring his theory of the formation of habits may help to clarify this. Skagestad, in discussing Peirce's theory of inquiry, points out:

> We come to philosophy with a network of preconceived opinions which it never occurs to us to doubt, because we are not aware that they can be doubted, or even that they are there. If we imagine that we doubt everything that can be doubted, we simply deceive ourselves and give our prejudices free play in determining our future beliefs. (Skagestad, 1981, 28)

Doubt cannot be in name only as is the case with those who take skepticisms as their starting point; doubt generated by choice is fictitious.

> It is important for the reader to satisfy himself that genuine doubt always has an external origin, usually from surprise; and that it is as impossible for a man to create in himself a genuine doubt by such an act of the will as would suffice to imagine the condition of a mathematical theorem, as it would be for him to give himself a genuine surprise by a simple act of the will. (SW.207, c.1905)

> A true doubt is accordingly a doubt which really interferes with the smooth working of the belief-habit. (CP5.510, c.1905)

Prejudices are taken-for-granted belief habits. For Peirce:

> A habit arises, when, having had the sensation of performing a certain act, *m*, on several occasions *a*, *b*, *c*, we come to do it upon every occurrence of the general event, *l*, of which *a*, *b* and *c* are special cases. That is to say, by the cognition that:
>   Every case of *a*, *b*, or *c*, is a case of *m*, is determined the cognition that
>   Every case of *l* is a case of *m*.
> Thus the formation of a habit is an induction and is therefore necessarily connected with attention or abstraction. Voluntary actions result from the sensations produced by habits. (Peirce, 1868)

Belief, he contended, has three properties:

> First, it is something that we are aware of; second, it appeases the irritation of doubt; and, third, it involves the establishment in our nature of a rule of action, or, say for short, a habit. As it appeases the irritation of doubt, which is the motive for thinking, thought relaxes and comes to rest for a moment when belief is reached. But, since belief is a rule for action, the application of which involves further doubt and further thought, at the same time that it is a stopping-place, it is also a new starting-place for thought. That is why I have permitted myself to call it thought at rest, although thought is essentially an action. The final upshot of thinking is the exercise of volition and of this thought no longer forms a part; but belief is only a stadium of mental action, an effect upon our nature due to thought, which will influence future thinking. (Peirce, 1878)

He stressed that

> belief is not a momentary mode of consciousness; it is a habit of mind essentially enduring for some time and mostly (at least) unconscious; and like other habits, it is (until it meets with some surprise that begins its dissolution)

perfectly self-satisfied. Doubt is of an altogether contrary genus. It is not a habit, but the privation of a habit. Now a privation of a habit, in order to be anything at all, must be a condition of erratic activity that in some way must get superseded by a habit. (SW.289, c.1905)

Living life through the medium of habits is supposedly unproblematic until one is, for one reason or another, assailed by doubt. As Peirce sees it: "The irritation of doubt causes a struggle to attain a state of belief. I shall term this struggle Inquiry" (CP5.374, 1893).

Some philosophers have imagined that to start an inquiry it was only necessary to utter a question whether orally or by setting it down upon paper and have even recommended us to begin our studies with questioning everything! But the mere putting of a proposition into the interrogative form does not stimulate the mind to any struggle after belief. There must be a real and living doubt and without this all discussion is idle. (Peirce, 1877)

In "The Fixation of Belief" (1877) Peirce could be said to be intent on provoking the "irritation of doubt" in those who have determined their beliefs without the benefit of logic. He is not so much concerned with individuals *qua* individuals but as members of a community. He said, in speaking of what he called the method of tenacity, that it "will be unable to hold its ground in practice."

The social impulse is against it. The man who adopts it will find that other men think differently from him and it will be apt to occur to him, in some saner moment, that their opinions are quite as good as his own and this will shake his confidence in his belief. This conception, that another man's thought or sentiment may be equivalent to one's own, is a distinctly new step and a highly important one. It arises from an impulse too strong in man to be suppressed, without danger of destroying the human species. Unless we make ourselves hermits, we shall necessarily influence each other's opinions; so that the problem becomes how to fix belief, not in the individual merely, but in the community. (Peirce, 1877)

Likewise, the second method he discussed, "the method of authority," takes into account the "social impulse." As Skagestad (1981: 33) puts it: "This method consists in letting the state legislate all beliefs, conduct systematic indoctrination, keep the population in ignorance of everything which may create doubt and punish all those who profess divergent beliefs." Equally, this method applies to religious dogma, political will, community acceptance, and peer pressure.

In failing to attend to experience, the third method, the *a priori* method, is related to the preceding two flawed methods. Peirce concedes:

This method is far more intellectual and respectable from the point of view of reason than either of the others. . . . But its failure has been the most manifest. It makes of inquiry something similar to the development of taste . . . and accordingly metaphysicians have never come to any fixed agreement, but the pendulum has swung backward and forward between a more material and a more spiritual philosophy, from the earliest times to the latest.

To satisfy our doubts, therefore, it is necessary that a method should be found by which our beliefs may be determined by nothing human, but by some external permanency—by something upon which our thinking has no effect. . . . Our external permanency would not be external, in our sense, if it was restricted in its influence to one individual. It must be something which affects, or might affect, every man. And, though these affections are necessarily as various as are individual conditions, yet the method must be such that the ultimate conclusion of every man shall be the same. (Peirce, 1877)

Notwithstanding that belief and doubt are not subject to will, deciding on a method for determining their validity is an act of volition. Furthermore, such a method requires the work of a community if it is to resist our eccentricities, the imposition of hegemony, or the vicissitudes of tastes. A method as choice is normative and arises in response not to the facts of the matter but to the interpretation of what is observed. As with phenomenology before it, and mathematics before that, observation is vital to inquiry, which in turn, is driven by purpose. What *is* observed vis-à-vis what is expected—what *should be* observable—is the aesthetics; the performance in response to any interpretation is not so much the action itself but rather what should be done in accord with aesthetics and is the ethics. Logic as normative entails concentrating on what are the possible consequences of the choices of actions in response to the aesthetic interpretation.

Aesthetics

Aesthetics as the normative science of values should not be confused with aesthetics as evaluation though they are of course related, the latter being dependent on the former for its principles. Within aesthetics, the *summum bonum* as ideal, is categorical; aesthetics as process is hypothetical by relationship. Thus, what we ought to *do*—where "do" includes "feel"—stands in relation to an end or ideal. According to Peirce:

The only satisfactory aim [ideal] is the broadest, highest and most general possible aim; and for any more definite information, as I conceive the matter, he has to refer us to the esthetician, whose business it is to say what is the state of things which is most admirable in itself regardless of any ulterior reason. (EP2. 253, 1903)

Aesthetics, as categorial, is a First; as a process, "esthetic quality is related to the three categories: It is Firstness that belongs to a Thirdness in its achievement of Secondness" (MS.301, 1903).

Let me illustrate this, holding in mind aesthetics' place within Peirce's classification of heuretic science, that is, discovery for the sake of discovery, philosophy, a positive, that is, experiential, heuretic science and following phenomenology, which focuses on the phaneron as it immediately presents itself and that, furthermore, the classifications and the categories which are revealed through phenomenology are formal.

Saying that the ideal of mathematics should be to produce conclusions that are elegant is to make a normative statement. Pronouncing the mathematical phenomenon, $m = E/c^2$ (Einstein's "second law"), as elegant is to make an aesthetic evaluation. While mathematics is superordinate to aesthetics and therefore is not dependent on it for its principles, the principles of aesthetics contribute to refinement of the principles and findings of mathematics.

Elegance may be defined as, inter alia, "the quality of ingenious simplicity and effectiveness." Remembering that defining anything does not assert its existence, elegance is a possibility; it is a First. Saying that $m = E/c^2$ is elegant is to actualize it through embodiment; it is a First of a Second.

This does not, however, warrant pronouncing elegance as the *summum bonum* of mathematics. Elegance as the ideal of mathematics is determined within aesthetics and is categorial. The steps in the process within Peirce's aesthetics are:

1. Physiology which investigates the *summum bonum* and discovers the physiology of phenomena in their Firstness;
2. Classification which inquires into the conditions of conformity to the ideal; and
3. Methodology which studies the principles governing the production of the aesthetic object.

Once again straying into metaphysics for a moment: nominating elegance as the ideal of mathematics may well be appropriate but to judge $m = E/c^2$ as elegant, therefore ideal, without consideration of its propositional value, is to opt for authority as the method of fixing belief. Such a method is not only nominalistic it also leaves unexamined the notion of elegance as an ideal. Furthermore, as a dogmatic statement, it blocks further inquiry. History tells us this is the antithesis of progress or process.

Joseph Raz (2003) speaks of the "practice of value," that is, that value depends on valuers. He delineates values and valuers and in so doing succeeds in separating the categorical process of determining the criteria and warrant of value from the normative process of valuing. This also serves to make Peirce's notion of aesthetics as valuing, intrinsic to the process of inquiry. This serves to emphasize continuation of the normative process from consideration of ideal as quality to ideal action.

Ethics

Peirce "takes the theory of the control of conduct and of action in general, so as to conform to an ideal, as being the mid-normative science" (CP1.573, 1905). In the same way that what he has to say on aesthetics should not be conflated to include aesthetic evaluation, so too ethics as a normative science should not be confused with practical ethics or pragmatics. Ethics as a normative science asks "to what end all effort shall be directed" (CP2.199, c.1903). Peirce argues, "The problem of ethics is to ascertain what end is possible" (CP5.134, 1903). In the same way it is important to differentiate Peirce's pragmatics from his ethics, so too is it important to differentiate ideals of conduct with motives to action.

> Every action has a motive; but an ideal only belongs to a line of conduct which is deliberate. To say that conduct is deliberate implies that each action, or each important action, is reviewed by the actor and his judgment is passed upon it, as to whether he wishes his future to be like that or not. His ideal is the kind of conduct which attracts him upon review. His self-criticism, followed by a more or less conscious resolution that in its turn excites a determination of his habit, will, with the aid of the *sequelæ*, modify a future action; but it will not generally be a moving cause to action. It is an almost purely passive liking for a way of doing whatever he may be moved to do . . . whether his own conduct or that of another person. (MS.283, 1906)

Contrary to what Descartes would have had us believe, blanket skepticism is not only *not* a necessary prelude to inquiry—it does not constitute a motive to action; to the contrary, it is an expression of "contempt prior to investigation." Genuine doubt, on the other hand, is, as discussed above, a motive to action—it is an itch that demands to be scratched.

Ethics inquires into the theory of the formation of habits of action that are consistent with a deliberately adopted ideal of conduct. Within Peirce's classification, the divisions of ethics are:

1. Physiology which discovers the ethical ideal and the physiology of conduct
2. Classification which inquires into the conditions of conformity to the ethical ideal

3. Methodology which inquires into the principles for actualizing the ethical ideal

Saying, for instance, that rigor is the right way of achieving mathematical elegance is to speak of ideals of conduct within mathematics. In his book *How Mathematicians Think* (2007) William Byers questioned rigor as the right way of conducting mathematics, arguing that it stymies the creativity necessary for the generation of original ideas. According to Douglas Hofstadter (1979), Euclid was the founder of rigor in mathematics and yet, as he is quoted in the OED following the definition of rigor as "strict accuracy, severe exactitude [*L15*]," "Euclid's lack of absolute rigor was the cause of some path-breaking mathematics." Creativity may well be argued as the right conduct for achieving the aesthetic ideal of pure mathematics. Not so in the case of applied mathematics where rigor, defined as "the strict application or enforcement of law, rule, etc. [LME]," is essential.

Whether these authors are right or wrong is not at issue here; my purpose in raising it is to point out that consideration of what is good conduct requires the normative science of ethics for exploring such propositions. It also illustrates that ethics is dependent on aesthetics, that is, what is considered as right action is dependent on what is conceived of as the *summum bonum*. In this respect ethics operates as a hypothetical imperative.

Logic

"Logic emerges in the scheme of the normative sciences as a special case of ethics, just as ethics involved a special determination of the aesthetic ideal" (Anderson 1995: 45). Peirce argued, "A logical reasoner is a reasoner who exercises great self-control in his intellectual operations; and therefore the logical good is simply a particular species of the morally good" (CP2.198, 1902-4). For Peirce "logic is a study of the means of attaining the end of thought" (Ibid.). He employs the term "logic" in two senses, which reveals his three divisions in his perennial classification of the sciences. Formatting this as it appears in the perennial classification of the sciences (and including the variety of labels he employed at different times in brackets) one can see that the narrow sense is folded into its broader sense:

Logic (also called general semeiotic; normative semeiotic; semiotic; semeiotic) studies the deliberate formation of habits of thought that are consistent with the logical end.

1. Speculative grammar: (on occasions referred to as Formal grammar; Hermeneutic; Logical syntax; Obsistent logic; *Stecheotic; Stechiology;*

Stoicheia; Universal grammar) inquires into the logical end and analyzes reasoning into ultimate components;

2. Critic: (or Critical logic; Logic; Originalian logic) studies the kinds and degrees of trust that are appropriate to different ways of reasoning; and
3. Methodeutic: (sometimes called Formal rhetoric; Heuristic; Method; Methodic; Methodology; Objective logic; Pure rhetoric; Speculative rhetoric; Transuasional logic) studies ways of pursuing different kinds of inquiry.

Logic, which inquires into the theory of the formation of habits of thought, where thought is mental action, is an extension of ethics. Mental action should not be conflated with mental processes which are the concern of empirical scientists such as psychologists and neuroscientists. As a normative science, logic is the theory of *conditions* of truth not what *is* true. Just as he was against logicism, so too was Peirce vocal in his opposition to the psychologism of such figures as Heinrich Christoph von Sigwart and John Stuart Mill.

In the same way, Peirce's semeiotic should not be confused with the semiology of the Swiss linguist, Ferdinand de Saussure whose *Cours de linguistique générale* (*Course in General Linguistics*) (1916) was published two years after Peirce's death. According to Floyd Merrell (2001):

> One of the chief distinctions between Peirce and Saussure lies in the scope of their theories. Peirce's semeiotics encompasses the range of all possible signs and their human and nonhuman makers and takers alike, regarding both inorganic and organic and living and nonliving domains—in addition to what is construed by dualists to be the realm of mind. This all-inclusive semeiotic sphere exists in stark contrast to Saussure's call for a "science of signs," which according to the proper conception was destined to become basically a "linguistic science," thus limited to distinctively human communication. (Merrell, 2001)

More importantly, however, James Liszka points out, "for Saussure, signs are primarily a psychological entity" (1996: 15) whereas "Peirce sees semeiotic as leading principles to sciences such as general and social psychology and linguistics." Peirce's semeiotic is normative; Saussure's semiology is empirical and as Liszka argues:

> The only way in which the logical or formal view of semeiotic and the empirical one would be compatible is if the empirical and the formal were treated the same. This is generally called the theory of psychologism; it is something Peirce argues fervently against. (Liszka, 1996: 15)

Speculative Grammar

Speculative grammar, the first division of Peirce's logic as semeiotic, is concerned with the formal conditions for "something" to count as a sign.

> A sign, or *representamen*, is something which stands to somebody for something in some respect or capacity. It addresses somebody, that is, creates in the mind of that person an equivalent sign, or perhaps a more developed sign. That sign which it creates I call the *interpretant* of the first sign. The sign stands for something, its *object*. It stands for that object, not in all respects, but in reference to a sort of idea, which I have sometimes called the *ground* of the representamen. (CP2.228, 1897)

Critic

"Whoever reasons *ipso facto* virtually holds a logical doctrine, his *logica utens*" (CP5.130, 1903). This is "in contradistinction to the result of the scientific study, which is called *logica docens* [and which] . . . is often considered to embrace the whole of logic; but a more correct designation is Critic" (CP2.204-205, 1902).

Critic, which "classifies arguments and determines the validity and degree of force of each kind" (CP1.191, 1903), is concerned with the formal conditions for counting a sign as true. While there is a range of forms of reasoning including analogy and metaphoric reasoning, Peirce articulated three kinds of arguments: abduction, deduction, and induction.

Methodeutic

The word "Methodeutic" does not appear in dictionaries so I assume it is one of Peirce's neologisms. He uses Methodeutic in preference to Method "as this word is also used in the concrete" (CP2.207, c.1902). It is my guess that he did not choose to name this division Methodology, which he also suggested in passing as an alternative to Method, because this is the label he used for a suborder in aesthetics and ethics and would be a subdivision of the divisions of logic. The word "methodology" came into English early in the eighteenth century to mean *the branch of knowledge that deals with method and its application in a particular field; the study of empirical research or the techniques employed in it.* By the mid-eighteenth century, it had also come to mean *a body of methods used in a particular branch of study or activity.*

> The whole discussion of the logical nature of the different kinds of possible signs makes up the first division of logic. . . . The second division . . . discusses the relation of signs to their objects, that is, their truth. The third division, Methodeutic, discusses the relations of signs to their interpretants, that is, their knowledge-producing value. (MS.793, c.1906)

I'd go further to say, their information produces value. In his unpublished manuscript, "Ideas, Stray or Stolen, about Scientific Writing," Peirce says "it is high time to acknowledge" that

> evidently our conception of rhetoric has got to be generalized; and while we are about it, why not remove the restriction of rhetoric to speech? What is the principal virtue ascribed to algebraical notation, if not the rhetorical virtue of perspicuity? Has not many a picture, many a sculpture the very same fault which in a poem we analyze as being too rhetorical. Let us cut short such objections by acknowledging at once, as an *ens in posse*, a universal art of rhetoric, which shall be the general secret of rendering signs effective, including under the term "sign" every picture, diagram, natural cry, pointing finger, wink, knot in one's handkerchief, memory, dream, fancy, concept, indication, token, symptom, letter, numeral, word, sentence, chapter, book, library and in short whatever, be it in the physical universe, be it in the world of thought, that, whether embodying an idea of any kind (and permit us throughout to use this term to cover purposes and feelings), or being connected with some existing object, or referring to future events through a general rule, causes something else, its interpreting sign, to be determined to a corresponding relation to the same idea, existing thing, or law . . . there is . . . a science to which should be referable the fundamental principles of everything like rhetoric, a speculative rhetoric, the science of the essential conditions under which a sign may determine an interpretant sign of itself and of whatever it signifies, or may, as a sign bring about a physical result. (MS.774, 1904)

He writes that "the methodeutic interest [is] in the devices which have to be employed to bring those new relations to light" (CP4.370, 1903). Tony Jappy's *Introduction to Peircean Visual Semeiotics* (2013) (his translation into English of his 2010 French edition) is what could be called a handbook for an increasingly visually oriented world. Poetry is another device, and one that for me can bring reality to light in ways that prosaic speech is incapable. Its purpose is not so much a literary way of communicating facts but as a device for reflecting experience which, being both temporal and concrete, enables understanding—sometimes indirectly as in my poem *Loss of confidence in the extent of my ignorance.*

When it comes to art, I'm one of the masses
who knows nothing about the subject;
who has no knowledge of art appreciation
but who knows what they like.
At least I thought that's what I was.

Because I like Turner's work
I go to the Tate when I'm in London.

Each time I leave none the wiser
as to how he did it—where he found
the pot of light with which he paints.

Today was one of those occasions.
I wandered from room to room,
ignorant, as usual, of what I was seeing
but nevertheless enjoying all that
I knew I liked—oblivious to what I didn't—

until I came across the room of
Recent 20th Century Acquisitions.
On the farthest wall I registered
a painting I didn't like.
and knew what I was seeing.

I went the length of that long room
to read the label and prove myself wrong;
to confirm I was right: that I knew nothing
about anything I was seeing—
beyond knowing what I liked.

I was wrong: I was right—
though I'd never heard of the artist;
nor recognised the place,
the painting was, like me, Australian.
I was shocked.

It took my childhood to realise I can't know everything;
my teens to admit I don't know it all;
my twenties to own up to not knowing some things;
my thirties to accept I don't need to know everything;
my forties to be at ease with the limitations of my knowledge.

This was something else.
It's one thing to know

Figure A1.4 Reimer's Interpretation of Aesthetic Experience. *Source*: Adaptation of Reimer (1970: 51).

what I don't know
but quite another to not know
what I do know.

What I realized many years later was what I had recognized in that painting as Australian: it was the light the artist had captured—the same light I'd grown up in—of southeast Australia. Peirce's semeiotics is not a language any more than is a musical notation or any other art form: their main purpose is not to communicate nor to express but rather to reflect or enlighten. Bennett Reimer, in the first edition of his book *A Philosophy of Music Education* (1970), shows how it works for art in a diagram.

Reimer has a clear understanding of the difference between emotion and feeling that is in accord with Peirce's explanations of feeling as the (a)esthetic ground of experience.

Liszka, in his paper "Peirce's new rhetoric" (2011: 439–449), recalls Peirce's "division of semeiotic, or the theory of signs into a trivium that mirrors the classical liberal arts: grammar, logic, and rhetoric."

> Semeiotically focused grammar, however, is concerned to study the essential features of a sign, "those conditions without the fulfilment of which [signs] would not be signs at all" (MS 1147A:111). Logic studies the conditions under which signs may represent their objects truthfully (see CP2.229, 1897), or "the conditions which determine reasoning to be secure" (CP2.1, 1902) (Liszka, 2011: 440)

He goes on to say:

> Although it is only developed programmatically, Peirce's rhetoric concerns the practice of inquiry and calls for an integration of rhetoric and logic on that basis, one which could possibly transform both disciplines in a fruitful way [...] Although these two divisions of semeiotic are well-developed, the same is not true of the third division of rhetoric. (Liszka, 439-440)

Better still is the realization that together with grammar they can inform. I could have given, for instance, a prosaic explanation of "truth in the long run" but to my mind none would bring greater clarity or be more likely to enlighten than Szymborska's poem π. Methodeutic is more than a means of what is commonly understood as communication. As the Thirdness of logic it is mediator, as Firstness of metaphysics it is creative, as degenerate Secondness of philosophy it is actual: "That which is met within the past, present or future" (Peirce, 1908). We all need to get out of silos to catch reality in passing.

The first two divisions of Peirce's logic are largely technical—not so his third division. Grammar discusses what is possible; critic considers what is admissible; Methodeutic can be said to focus on what is advantageous. Because its aim is truth and truth is what it is regardless of you or I, "advantageous" should not be taken to mean "expedient" or merely "instrumental." These constitute rationalizations. To illustrate where rationalization can lead, let me relate a piece of facetious humor by Michael Kindler, Faculty of Education, University of Western Sydney, I found some years ago in the otherwise mundane machinations of the University of Sydney Senate papers.

Schubert's Unfinished Symphony No. 8, *Die Unvollendete*

An economic parable for all those concerned with the effective use of resources in times of educational restructuring and financial stringency.

A Management Review Task Force in a new cluster visited an outer Western Sydney high school to conduct an educational audit. The visit coincided with one of the concerts of the Sydney Symphony Orchestra, to which the Principal was in the habit of going. On this occasion, because he had to attend a global budget meeting followed by a staff selection committee, he could not go. With his usual generosity, however, he gave his ticket to the leader of the Management Review Task Force, who had never been to a symphony concert before. The main work that night was Shubert's *Unfinished Symphony*.

When he asked his visitor the following morning how he had enjoyed the concert, the Principal was surprised to be handed a typewritten report:

1. *For considerable periods the four oboe players had nothing to do. The number should be reduced and their work be more conveniently spread over the whole concert, thus eliminating peaks of activity.*
2. *All the 12 violins were playing identical notes. This seems unnecessary duplication. The staff of this section should be drastically cut and if a large volume of sound is really required, this could be obtained by means of an electronic amplifier.*
3. *Much effort was absorbed in the playing of demi-semiquavers. This seems to us an excessive refinement and it is recommended that all notes be rounded up to the nearest semiquaver. If this were done it should be possible to use trainees and lower grade operators.*
4. *There seems to be too much repetition of some musical passages. No useful purpose is served by repeating with horns the passage already handled by the strings. If all such redundant passages were eliminated, the whole concert time of two hours would have been reduced to twenty minutes and there would have been no need for an interval.*

If the composer had attended to these matters, he would probably been able to finish his symphony.

Methodeutic as being concerned with the relations of signs to their interpretants leads to consideration of Peirce's Pragmatic Maxim and his pragmaticism. Peirce, in correspondence with William James, explained to his friend the role of logic in pragmatism and of the Interpretant in logic as semeiotic.

> The Sign creates something in the Mind of the Interpreter, which something, in that it has been so created by the sign, has been, in a mediate and relative way, also created by the Object of the Sign, although the Object is essentially other than the Sign. And this creature of the sign is called the Interpretant. (EP2:493-4, 1909)

The Interpretant: Empiricists would have it that there is a one-to-one correspondence between perception and interpretation. Peirce says they are mistaken; that this is conjecture on their part. This is what he called the Immediate Interpretant which "consists in the Quality of the Impression that a sign is fit to produce, not to any actual reaction" (CP8.315, 1909). In a letter to Lady Welby with whom he corresponded over a number of years, he described the three forms of the Interpretant as representation.

> I understand the [Immediate Interpretant] to be the total unanalyzed effect that the Sign is calculated to produce; and I have been accustomed to identify this with the effect the sign first produces or may produce upon a mind, without any reflection upon it. . . . I might describe my Immediate Interpretation, as so much of a Sign that would enable a person to say whether or not the Sign was applicable to anything concerning which that person had sufficient acquaintance. . . . My Immediate Interpretant is implied in the fact that each Sign must have its peculiar Interpretability before it gets any Interpreter. . . . The Immediate Interpretant is an abstraction, consisting in a Possibility.
> My Dynamical Interpretant consists in direct effect actually produced by a Sign upon an Interpreter of it. . . . My Dynamical Interpretant is that which is experienced in each act of Interpretation and is different in each from that of the other. . . . The Dynamical Interpretant is a single actual event.
> My Final Interpretant is . . . the effect the Sign would produce upon any mind upon which the circumstances should permit it to work out its full effect. . . . [It] is the one Interpretative result to which every Interpreter is destined to come if the Sign is sufficiently considered. . . . The Final Interpretant is that toward which the actual tends. (Letter to Lady Welby, SS.110-1, 1909)

At one point, rather than describe processes as teleological which suggests purpose, Peirce coined the neologism, "finious," "to express their tendency toward a final state" (CP7.471, 1898).

Hulswit (1997, 2002) describes this process as *causation* and distinguishes it from the classical theories of *causality*. Discussion of Peirce's theory of causation might be considered more relevant to his metaphysics, but raising it here serves three purposes. First it helps to clarify Peirce's Interpretant; second, it illustrates the natural progression of his classification; and third, it brings into context his Pragmatic Maxim. Peirce first enunciated this maxim in 1873 and reiterated it in 1905 in his discussion of the difference between what he came to call pragmaticism—the pragmatism he authored—and the popularized versions of pragmatism.

> Pragmaticism was originally enounced in the form of a maxim, as follows: Consider what effects that might conceivably have practical bearings you conceive the objects of your conception to have. Then, your conception of those effects is the whole of your conception of the object. (SW. 289, 1905)

Hulswit points out that causation "assumes causal relata to be discrete entities between which there is not even a hint of continuity" (2002: 176) and furthermore, regarding the causal relata, theories of causality are ambiguous. He argues that

> though most contemporary philosophers hold that the causal relata are *events*, there are also some philosophers that hold that they are *facts* and again some who hold that, next to event causality or fact causality, there is also agent causality in which the agents are conceived as *substances*. . . . [I]t would appear that the received view's apparent insistence on events is trapped between Aristotle's substance ontology and the modern scientific fact ontology. (Hulswit, 176 & 178)

Through his study of the history of the concept of cause, Peirce observed these discrepancies, which he identified as being between (1) the Aristotelian conception, (2) the modern physicist's conception, and (3) the currently accepted view (RLT.197-202, 1898). The reconciliation led to Peirce's theory of causation, most clearly enunciated in his 1902 paper *On science and natural classes* (EP2, item 9). Hulswit gives a synopsis:

> Peirce developed the highly original view that each act of causation involves an efficient component, a final component and a chance component. The efficient aspect of causation is that each event or fact is produced by a previous event or fact (the efficient cause). The teleological aspect is that each event is part of a chain of events with a definite tendency. The chance component is that each event has some aspect that is determined neither by the efficient nor by the final cause. (Hulswit, 2002: 187)

Understanding this process of causation gives an inkling of what Peirce meant by "truth in the long run." Szymborska told us in her poem π [pi]

The series comprising π
doesn't stop at the edge of the sheet

Likewise, the finious process of causation perseveres. As Peirce points out:

> Experience can only mean the total cognitive result of living and includes inter-
> pretations quite as truly as it does the matter of sense. Even more truly, since this
> matter of sense is a hypothetical something which we never can seize as such,
> free from all interpretive working over. (CP7.538, n.d.)

In terms of causation, the universe should be considered as living and any
definition of life must take this into account. To this end, I suggest that power
could be considered as the capacity to cause and effect. This does not define
power any more than "never having to say you're sorry" defines "love."
These are matters for metaphysics. While observation alone may serve epistemology, semeiotic is essential for developing a metaphysics that is sound
and which can lead to discovery in the special sciences.

Peirce's *Speculative Rhetoric/Methodeutic* is wide ranging. Liszka (2011:
440) recognizing this, covers the broad territory in part, noting:

> however it is called throughout its history, it is considered in a number of
> different ways: as the matter of conducting research wisely (MS165, 1895),
> or as how truth must be properly investigated (MS320:27, 1907; MS606: 15,
> 1905, CP1.191, 1903), as the formal conditions for the attainment of truth
> (CP2.207, 1902); the ordering and arranging of inquiries (MS478, 1903;
> MS452:9, 1903; CP3.430, 1896; CP2.106-110, 1902), the study of the general
> conditions under which a problem presents itself for solution (CP3.430, 1896),
> the method of methods (CP2.108, 1902), the management and economy of
> hypotheses (MSL75, 1902). But it also has to do with the power of symbols to
> appeal to a mind (CP4.116, 1893; CP1.559, 1867; CP1.444, 1896), or condi-
> tions for the intelligibility of symbols (MS 340: 34, 1864-5; Wl:1175, 1865;
> MS 774: 9-11, 1904), or the clarity of ideas (MSL75, 1902; MS322:12, 1907);
> it is concerned with the transmission of ideas (CP1.445, 1896; CP2.93, 1902),
> the consequences of accepting beliefs (NEM4: 291), or how to render signs
> effective (MS74: 2, 1904). In addition, speculative rhetoric studies the growth
> of Reason (NEM4: 30-31), the science of the general laws of a symbol's
> relation to other systems of symbols (W1: 258, 1865), evolution of thought
> (CP2.108, 1902; CP2.111, 1902), the advancement of knowledge (MS 449:
> 56, 1903) and the influence of ideas (NEM4: 31); it is concerned with system-
> atic and architectonic matters (MS346: 3, 1864-5; CP4.116, 1893). (Liszka,
> 2011: 440)

In his *Minute Logic* (CP2.105, 1902) Peirce wrote: "All this brings us close to Methodeutic, or Speculative Rhetoric. The practical want of a good treatment of this subject is acute." He had intended to write such a book but did not, largely because he was denied the Carnegie grant which would have funded his work. The practical want of a good treatment of this division, of a suborder, of an order, of a class, of a branch, remains acute. Tony Jappy (2013) is one who demonstrates the observable operation of Peirce's semeiotic, especially his speculative rhetoric in relation to abduction and metaphor.

Ultimately, it is not logic in the atemporal classical sense but rather in the temporal sense of his semeiotic that sets Peirce apart from other pragmatists in particular and other philosophers in general. The logicist position of analytic philosophy creates an irresolvable barrier to the inclusion of metaphysics in systematic inquiry. For the rationalists, logic is the be-all and end-all of philosophy—metaphysics is *merely* recreation. From the reductionist empiricists' point of view, metaphysics is irrelevant. To many modern-day specialist scientists, metaphysics is "stuff and nonsense." Peirce's method reconciled the rationalist and empiricist positions, and even though he had a low opinion of much of enlightenment metaphysics, he came to see metaphysics as essential to inquiry.

METAPHYSICS

In the production of knowledge, mathematics produces necessary inferences. Nevertheless, in confining itself, as it does, to drawing necessary conclusions from entirely hypothetical constructions, without caring in the least whether these constructions apply to anything real, it does not produce positive knowledge. The discovery of positive facts begins with philosophy and is of two kinds: perceptual and conceptual knowledge. The knowledge of phenomenology is perceptual knowledge which Peirce spoke of as follows: "The knowledge which you are compelled to admit is that knowledge which is directly forced upon you and which there is no criticizing, because it is directly forced upon you" (CP2.141, c.1902). "Perceptual knowledge, although in some sense general, is beyond human control" (Curley, 1969: 92–3). For Peirce "a classification of arguments, antecedent to any systematic study of the subject, is called the reasoner's *logica utens*, in contradistinction to the result of the scientific study, which is called *logica docens*" (CP2.204, 1901). It is through *logica docens* that conceptual knowledge, which concerns criticism and control, is developed and clarified. According to Peirce:

Logic requires that the more abstract sciences should be developed earlier than the more concrete ones. For the more concrete sciences require as fundamental principles the results of the more abstract sciences, while the latter only make use of the results of the former as data; and if one fact is wanting, some other will generally serve to support the same generalization. (CP6.1, 1898)

Such criticism arises because, largely as a result of nominalism, there is a general misunderstanding of what is meant by "facts." This misunderstanding is not so much to do with a distinction between facts and fiction, as it is with their temporal ontological status.

Ontology

Taking two definitions of metaphysics: (1) the branch of philosophy that studies the ultimate structure and constitution of *reality* and (2) the branch of philosophy concerned with the ultimate nature of *existence*, Plato is more readily equated with the former and Aristotle with the latter. Peirce was a realist—an extreme realist—or as Rosa Mayorga (2007: 152) argues, a scholastic "realicist." She summarizes her theory:

> Peirce incorporates elements of nominalism, idealism and scholastic realism in his notion of the real. He accepts the nominalist notion that generals, or universals, are of the nature of thought, but rejects that doctrine's claim that only individuals are real. He accepts the idealist notion that reality is relative to the mind, but rejects Berkeley's description of reality. He accepts the scholastic notion of the reality of universals and adapts Scotus's *realitas* to reflect his own notion of reality, but rejects Scotus's notion of contraction. Peirce then takes all these elements, adds some of his own such as synechism and combines them

	PLATO	ARISTOTLE	REALISTS	NOMINALIST	PEIRCE
UNIVERSALS	Exist Real	Exist (?) Real (maybe)	Don't Exist Real	Don't Exist Not Real	Don't Exist Real
SINGULARS	Exist Not Real	Exist Real	Exist Real	Exist Real	Exist No Reality

Figure A1.5 Table Showing Peirce's Incorporation of Elements of Nominalism, Idealism, and Scholastic Realism Together which with Elements of His Such as Synechism Suggesting his Theory of Realism Called Here Scholastic Realicism. *Source*: Adaptation of Mayorga (2007: 152).

into his own theory, which I have suggested should be called his scholastic realicism.

This may make more sense if we understand what Boler (1963: 138–143) points out and Moore (PM. xxxix, 2010) brings to our attention that

> Peirce tends to drain Seconds of their content and restrict the reality of Secondness to instantaneous; what we ordinarily count as an individual—an individual person say—is not a Second but rather a Third, a law that governs the host of Seconds that go to make up the person. (See Peirce EP2.221-222, 1903)

For Peirce, being is a monad, existence a dyad; only a triad is real.

He was also an antifoundationalist, but as Richard Bernstein (2013) notes: "Peirce realized that in criticizing foundationalism he was attacking many of the most cherished doctrines and dogmas that constituted modern philosophy." Unlike modern-day scientists, however, he did not see this as obviating metaphysics—whether we recognize it or not, we all have a metaphysical position; our metaphysics is the fabric of our everyday reality. He concluded that metaphysics needed a total overhaul. Having pronounced radical skepticism untenable, he was not blind to the inadequacy of either empiricism or rationalism on their own for discovering truth. Exploration needs to begin with creativity, imagination, a theory, a hypothesis, empiricism, and rationalism; it needs semeiotic.

Peirce railed against all dogmatists, be they philosophers, teachers, scientists, or theologians. Let me point out here that we are unlikely to know what is impossible until we know the full breadth and depth of possibility, which, to quote a popular song of the 1950s, will not be "until the twelfth of never."

Religious Metaphysics

Because the use of the words "religion" and "religious" generates deep suspicion and antagonism I will defer to Paul Tillich's (1957) definition of "faith" as "the state of being ultimately concerned" and "religion" broadly as "ultimate concern." Tillich did not exclude atheists in his exposition of faith, for as he saw it, "even if the act of faith includes the denial of God, where there is ultimate concern, God can be denied only in the name of God" (1957: 52). Peirce's view of religion was that, in the individual, it was "a deep recognition of a something in the circumambient All." However, on his view of reality, he saw that "religion cannot reside in its totality in a single individual. Like every species of reality, it is essentially a social, a public affair" (CP6.428, 1898). His view was that "religion is a life and can be identified with a belief only provided that belief be a living belief—a thing to be lived rather than said or thought" (CP6.439, 1893).

Although he was Christian, he did not push any doctrinal line. In his paper "Answers to Questions about My Belief in God" (MS.845, 1906) he says:

> "God" is a vernacular word and, like all such words, but more than almost any, is vague. No words are so well understood as vernacular words, in one way; yet they are invariably vague; and of many of them it is true that, let the logician do his best to substitute precise equivalents in their places, still the vernacular words alone, for all their vagueness, answer the principal purposes. This is emphatically the case with the very vague word "God," which is not made less vague by saying that it imports "infinity," etc., since those attributes are at least as vague. (PWP.275, 1906)

Furthermore, he is consistent with his radical realism and speaks not of the *existence* but of the *reality* of God, as in his paper "A Neglected Argument for the Reality of God" (Peirce, 1908). Anderson points out in his commentary on the paper:

> The actual, or the existent, on the other hand, is one moment of the real, the reality of Secondness. The actual is thus that which is located in causal relations; it "is that which is met in the past, present, or future." The importance of the distinction here is that Peirce wants to discuss the "reality" of God, not the "actuality" or the "existence" of God under the narrower conception. (1995: 141)

I find the word "located" to be the key here: that which can be located *exists*; not so that which is *real*. The difference would confound those nominalists who, in the face of all and any of the proofs to the *existence* of God (by any name), speak of themselves as agnostic. To call oneself an agnostic is not necessarily an expression of disbelief in God but rather to say that one cannot *know* of the existence or otherwise of God.

Peirce's outlook on religion developed in relation to his view of science. For him, the purpose of religion "is to conduct our lives . . . in such a way as to ameliorate human existence" (Anderson 2002, p. 4). Science, on the other hand, is an inquiry which is systematically directed toward developing ideas and discovering truth. Anderson points out that Peirce

> envisioned religion in a reciprocal dependence with science; the two must engage in an ongoing dialectical relationship. An idea that is effective as a religious belief, if it is to be theorized about, must turn itself over to scientific inquiry, to criticism. (Ibid. p. 7)

Peirce was scathing in his condemnation of theologians. Because his tirades against theology were so vitriolic, rather than quote him directly, let me cite in part Anderson's synopsis of his argument (ibid. p. 6):

> Theology's method, if it is a method, is to express and defend tenaciously and authoritatively some specific version of religious ideas. To do this, it tries to specify the ideas so particular rules and interpretations can be nailed down. For example, the vernacular God is replaced with a named being or beings who are historically located, embodied, or otherwise definitely described. Likewise, the good is reduced to a narrow formula of behaviour, a set of rules that curtails human variety and flexibility in dealing with life situations. In short, theologians produce and defend creeds and doctrines.

Physical Metaphysics—Cosmology

Peirce's cosmogony begins with his notion of boundless freedom.

> The initial condition, before the universe existed, was not a state of pure abstract being. On the contrary it was a state of just nothing at all, not even a state of emptiness, for even emptiness is something. (CP6.215, 1898)

So too is time for something, being sequenced and therefore a beginning is also excluded.

> We start, then, with nothing, pure zero. But this is not the nothing of negation. For not means other than and other is merely a synonym of the ordinal numeral second. As such it implies a first; while the present pure zero is prior to every first. The nothing of negation is the nothing of death, which comes second to, or after, everything. But this pure zero is the nothing of not having been born. There is no individual thing, no compulsion, outward nor inward, no law. It is the germinal nothing, in which the whole universe is involved or foreshadowed. As such, it is absolutely undefined and unlimited possibility—boundless possibility. There is no compulsion and no law. It is boundless freedom.
>   So of potential being there was in that initial state no lack. (CP6.217, 1898)

Peirce argued that nothing necessarily, that is, deductively, results from the nothing of boundless freedom. From the perspective of Firstness, potentiality became potentiality of this or that sort—that is, of some quality.

> Thus the zero of bare possibility, by evolutionary logic, leapt into the unit of some quality. This was hypothetic inference. Its form was:
>   Something is possible,
>   red is something;
>   ∴ Red is possible. (CP6.220, 1898)

Responding philosophically to the question: "Why there is something rather than nothing?" and to the idea of an evolving, developing telos (without the need for recourse to, or denial of a Creator) becomes more straightforward when approached via Peirce's categories. First possibility, Second actuality, and Third probability. First nothing which is teaming with all possibility; Second chaos, which is a First of a Second—unstable, disorganized, unreasonable "something"; Third Reality: from infinite possibility, emerges an arbitrary, disorderly rabble, which "finds" rhyme and reason and value, and voilà: the Cosmos. From nothing, something. As Peirce said:

> Without going into other important questions of philosophical architectonic, we can readily foresee what sort of a metaphysics would appropriately be constructed from those conceptions. Like some of the most ancient and some of the most recent speculations it would be a Cosmogonic Philosophy. It would suppose that in the beginning—infinitely remote—there was a chaos of unpersonalized feeling, which being without connection or regularity would properly be without existence. This feeling, sporting here and there in pure arbitrariness, would have started the germ of a generalizing tendency. Its other sportings would be evanescent, but this would have a growing virtue. Thus, the tendency to habit would be started; and from this, with the other principles of evolution, all the regularities of the universe would be evolved. At any time, however, an element of pure chance survives and will remain until the world becomes an absolutely perfect, rational and symmetrical system, in which mind is at last crystallized in the infinitely distant future. (Peirce, 1891)

A driving force of his cosmogony and cosmology was his response to and rejection of popularly held beliefs in the nineteenth century, in agnosticism, necessarianism, and mechanism. Agnosticism is here used to express the notion of "unknowable" which may well have been promulgated to keep religion off limits to the probing of science. The closely related necessarianism and mechanism he saw as uncongenial to any possibility of freedom or novelty in the universe. As he saw it:

> The only possible way of accounting for the laws of nature and for uniformity in general is to suppose them results of evolution. This supposes them not to be absolute, not to be obeyed precisely. It makes an element of indeterminacy, spontaneity, or absolute chance in nature. (Peirce, 1891)

> That idea has been worked out by me with elaboration. It accounts for the main features of the universe as we know it—the characters of time, space, matter, force, gravitation, electricity, etc. It predicts many more things which new observations can alone bring to the test. May some future student go over

this ground again and have the leisure to give his results to the world. (Peirce, 1891)

Keeping in mind that Peirce's cosmology is not wild speculation but is principle dependent on all the preceding classes in his perennial classification and in turn provides data to those classes, its purpose is to develop the concepts for special science to test. In his cosmology these concepts are what he calls *synechism, tychism*, and *agapasm*, with, according to Esposito (2001), synechism as the keystone of Peirce's metaphysics.

> The term *synechism* is derived from the Greek *syneche*, meaning "continuity," or "held together." As a methodological doctrine, synechism exhorts us to attempt to tie together all known facts about the universe, leaving no loose ends. . . . *Tychism* (from the Greek *tyche*, meaning "chance") is the hypothesis that the world is essentially indeterministic, that no law of nature is absolutely exact. . . . *Agapasm* (from the Greek *agape*, meaning "love") posits the reality of final causes in the processes of the world. . . . As in the cosmologies of some of the ancient Greeks, love is understood here as a uniting or attractive force that draws all the component parts of the universe into a coherent whole. (Reynolds, 2002: 11)

Those who reject the role of metaphysics in the furtherance of scientific discovery are perhaps rejecting, as did Peirce, the shambolic state of much of what has passed itself off as metaphysics. Peirce's cosmology was not idle speculation but was informed downward from mathematics and upward from evolutionary sciences. Listed next are just a few of these.

Downward:

- Pierre Varignon [1654–1722], the advocate of infinitesimal calculus, who recognized the importance of a test for the convergence of series, simplified the proofs of many propositions in mechanics, adapted the calculus of Gottfried Leibniz [1646–1716] to the inertial mechanics of Newton's Principia and treated mechanics in terms of the composition of forces.
- Jacob Bernoulli [1654–1705] who described the known results in probability theory and in enumeration, often providing alternative proofs of known results and who introduced the theorem known as the law of large numbers.
- Thomas Bayes [1702–1761] whose theorem proposes that evidence confirms the likelihood of a hypothesis only to the degree that the appearance of this evidence would be more probable with the assumption of the hypothesis than without it.
- Joseph Lagrange [1736–1813], one of the creators of the calculus of variations, deriving the Euler–Lagrange equations for *extrema* of functionals. He also extended the method to take into account possible constraints, arriving

at the method known as Lagrange multipliers, and invented the method of solving differential equations known as variation of parameters, applied differential calculus to the theory of probabilities and attained notable work on the solution of equations. At a later period, he reverted to the use of infinitesimals in preference to founding the differential calculus on the study of algebraic forms.

- Augustin-Louis Cauchy [1789–1857] who initiated the project of formulating and proving the theorems of infinitesimal calculus in a rigorous manner.
- Nikolai Lobachevski [1792–1856] who developed a non-Euclidean geometry, denying the truth of Euclid's parallel postulate by supposing that there may be two or more such lines passing through a given point.
- Augustus De Morgan [1806–1871] who recognized the need to expand the notion of logical validity beyond the narrow confines of Aristotelian syllogistic and developed the standard statement of De Morgan's Theorems, a pair of logical relationships.
- Hermann Grassmann [1809–1877] who developed the idea of an algebra in which the symbols representing geometric entities such as points, lines, and planes are manipulated using certain rules. He represented subspaces of a space by coordinates leading to point mapping of an algebraic manifold now called the Grassmannian.
- George Boole [1815–1864] who developed a symbolic system for the expression and evaluation of categorical syllogisms, understood as elements in the logic of classes.
- Bernhard Riemann [1826–1866] who developed field theory as a mathematical description of phenomena as apparently diverse as gravitation, magnetism, electricity, and light and contributed to the development of topology and non-Euclidean geometry.
- John Venn [1834–1923] who applied the insights of Boole, Euler, and others in developing a diagrammatic method for testing the validity of categorical syllogisms and who contributed to the development of modern theories of probability.
- Karl Pearson [1837–1936] who established the discipline of mathematical statistics, and who asserted that the laws of nature are relative to the perceptive ability of the observer.
- Georg Cantor [1845–1918] who developed a modern set theory as the foundation for all of mathematics and used the "diagonal proof" to demonstrate that lines, planes, and spaces must all contain a non-denumerable infinity of points, that is, they cannot be counted in a one-to-one correspondence with the rational numbers. The reality of trans-finite quantities within the set of real numbers lead in turn to "Cantor's paradox"—that every set has more subsets than members, so that there can be no set of all sets.

Downward and upward:

- James Clerk Maxwell [1831–1879] whose work in producing a unified model of electromagnetism is considered to be one of the greatest advances in physics. He also developed the Maxwell distribution, a statistical means of describing aspects of the kinetic theory of gases. These two discoveries helped usher in the era of modern physics, laying the foundation for future work in such fields as special relativity and quantum mechanics.

Upward:

- Jean-Baptiste Lamarck [1744–1829] whose study of invertebrates led to the conviction that species evolve through the hereditary transmission of acquired traits, by means of which species perfect their adaptation to their environment in an optimal fashion. His contribution to evolutionary theory consisted of the first truly cohesive theory of evolution, in which an alchemical complexifying force drove organisms up a ladder of complexity and a second environmental force adapted them to local environments through the use and disuse of characteristics, differentiating them from other organisms.
- George Bentham [1800–1884] who is quoted as saying: "We cannot form an idea of a species from a single individual, nor of a genus from a single one of its species. We can no more set up a typical species than a typical individual."
- Clarence King [1842–1901] who publicly put forth arguments against the followers of Lyell and Darwin. He was not trying to completely return to the catastrophic viewpoint, but he felt the uniformitarians were going too far, as he had personally observed that the rate of change had not remained steady. His argument would be verified by the power laws of ubiquity.

And of course, Charles Darwin [1809–1882] whose work Peirce discussed at length with Chauncey Wright [1830–1875] cofounder with Peirce of The Metaphysical Club and whose views on Darwinism played a significant role in shaping the ideas of the other members of the club.

An aspect of Peirce's cosmology was his doctrine of matter:

> The one intelligible theory of the universe is that of objective idealism, that matter is effete mind, inveterate habits becoming physical laws. But before this can be accepted it must show itself capable of explaining the tri-dimensionality of space, the laws of motion and the general characteristics of the universe, with

mathematical clearness and precision; for no less should be demanded of every philosophy. (Peirce, 1891 "The architecture of theories.")

My concern, though, is not with matter *per se* but rather, with mattering.

THE SPECIAL SCIENCES

The special sciences constitute the third heuretic science, that is, their concern is with discovery for the sake of discovery. According to Peirce, "Sciences must be classified according to the peculiar means of observation they employ" (CP1.101, c. 1896). Every discovery has its origin in an observation—observation is the foundation on which all the sciences are erected—in the case of the Special Sciences, observations are of previously unknown phenomena and are empirical.

Peirce identified two sub-classes of the special sciences: physical sciences and humanistic sciences and the same three orders for each.

1. Nomological sciences: study the ubiquitous phenomena of the physical and psychical universes, ascertain their general laws, and measure the quantities involved.
2. Classificatory sciences: describe and classify the various kinds among the objects studied and endeavor to explain them by means of the general laws.
3. Explanatory sciences: study and minutely describe individual objects and events and subsequently seek to explain using the findings of the nomological and classificatory sciences.

As with the preceding classes in the perennial classification, the special sciences are principle dependent and provide data upward that can serve to refine the principles of the sciences superordinate to them.

In his book *What Is this Thing Called Science?* (1982) Alan Chalmers rejects the popularly held belief of science as *product*. Science as *inquiry* is a process. Chalmers speaks of the *aim* of science. In arguing for the objectivity of observation in science, he attempts to capture the middle ground between the extremes of the universal method and skeptical relativism by appealing to historically contingent standards implicit in successful practices. He rejects inductivism, which he argues is a mistaken attempt to formalize the popular view of science, and then Popper's falsificationist account says that attention to history strongly suggests that both it and the inductivist accounts are too piecemeal.

Neither the naive inductivist emphasis on the inductive derivation of theories from observation, nor the falsificationist scheme of conjectures and falsifications, is capable of yielding an adequate characterization of the genesis and growth of realistically complex theories. (Chalmers 1982: 77)

Following this he makes summaries of two attempts to analyze theories as organized structures: Lakatos's research programs and Kuhn's paradigms, which have given rise to a debate concerning the two contrasting positions associated with them—"rationalism" and "relativism." He sums up the discussion by noting, "Lakatos aimed to give a rationalist account of science but failed, whilst Kuhn denied that he aimed to give a relativist account of science but gave one nevertheless" (ibid. 108).

He proposes "a way of analyzing science, its aims and its mode of progress, which focuses on features of science itself, irrespective of what individuals or groups might think" (ibid. 110). In arguing for the objectivity of observation in science, Chalmers in his *Science and Its Fabrication* (1990) stresses that he is arguing against skeptical relativism, not fallibilism. His arguments stand against those of empiricists and instrumentalists who put forward the notions that in being theory dependent, science is on the one hand subjective and on the other unable to establish truth. He counters these by arguing that

if we interpret "objective" to mean something like "testable by routine procedures" . . . what is correct about the "theory dependence of observation" thesis is not that observation in science lacks objectivity, but that the adequacy and relevance of observation reports within science is subject to revision. Observation in science may be objectified, but we do not thereby have access to secure foundations for science. (Ibid. 59)

Jaime Nubiola, in his paper "The Branching of Science According to C. S. Peirce" (1995), argues, "the key to the advancement of knowledge and to the development of sciences is not revolution, but growth in a very peculiar mixture of continuity and fallibilism."

In order to make sense of our world, it is sometimes necessary to backtrack through the history of thought to discover where we may have taken a wrong turn. In Australia, on those occasions when a highway's egress and exit are side-by-side, motorists who mistakenly enter an exit are informed by a sign, a few meters into the slipway: "Go back you are going the wrong way." Peirce, in going back, identified the major culprit as the nominalism of William of Ockham and reinstated and improved upon Duns Scotus's realism.

Likewise, Jon Ogborn and Edwin F. Taylor trace the transition from fundamental quantum mechanics to derived classical mechanics in their paper "Quantum Physics Explains Newton's Laws of Motion" (2005):

Newton was obliged to give his laws of motion as fundamental axioms. But today we know that the quantum world is fundamental and Newton's laws can be seen as *consequences* of fundamental quantum laws. (Ogborn & Taylor, 2005)

Werner Heisenberg, in his 1927 paper that introduced the uncertainty principle to the world, established that there are pairs of quantities in the quantum world that cannot both be measured to an arbitrary level of precision at the same time.

Many theoretical and experimental physicists have worked over the years to grasp uncertainty. Translating insights from the theory of information devised by the American mathematician Claude Shannon to the quantum world, the Dutch physicists Hans Maassen and Jos Uffink showed in 1988 how it is impossible to reduce the Shannon entropy associated with any measurable quantum quantity to zero and that the more you squeeze the entropy of one variable, the more the entropy of the other increases. Information that a quantum system gives with one hand, it takes with the other. To the question: "What is it that keeps quantum theory as weird as it is and no weirder?" Stefanie Wehner and Jonathan Oppenheim's answer is (*Science* 2010 v.330: 1072): the uncertainty principle (Anathaswamy, 2011).

Parmenides is the philosopher for those focused on atemporal equilibrium physics where nothing changes. Heraclites would get the nod from non-equilibrium physicists: those who believe history matters and who are concerned with the physics of complex systems. Peirce had great respect for the genius of the likes of Newton and Darwin but he did not share their fatalist or a priori assumptions about natural law for the former and mechanistic evolution for the latter. Peirce did not refute Darwin, for instance, but he was of the view that his theory was inadequate to explain reality. He considered a combination of the ideas of Darwin with those of Clarence King [1842–1901] the American geologist and author of the paper "Catastrophism and the Evolution of Environment" (*American Naturalist*, 11, 8: 449–470, 1877) and Jean-Baptiste Lamarck [1744–1829] the French evolutionist, more fitting and intelligible. Peirce's pragmaticism is a theory of meaning. Although he saw truth as something to be sought in the long run, his method of inquiry focused on the discovery of meaning.

Those special scientists, who reject philosophy out of hand, do themselves a great disservice in the pursuit of their particular field. I am talking here, not of knowing but of *doing* philosophy. Granted, there is much that calls itself philosophy—or part thereof—that may be considered of dubious merit, but competence in doing philosophy enables one to identify the charlatans. Dogmatism is one such, but then, so too is skepticism if it is not genuine.

Every aspect of that which is called the scientific method is the product of philosophy. Picking-up methods by rote and running with them are limiting. Take, for example, diagnosis in medicine: Donald Stanley notes in "The Logic of Medical Diagnosis" (Stanley & Campos, 2013) research has shown that "diagnostic error accounts for 40,000 to 80,000 deaths per year" and asks, "Why does improving diagnosis go largely under-mentioned, subservient to evaluating evidence for treatment decisions? Why have medical school educators devoted more time to research on treatment than to diagnosis?" (ibid. 301). Stanley's co-author, Campos, points to a poor understanding of abduction.

> When we are confronted with new, often puzzling, facts that we seek to explain, we are in a situation that requires us to make a conjecture that would explain the facts and to adopt the conjecture provisionally as a hypothesis that we may test. The abductive suggestion consists in the conjecture that a general rule—say the general character of a certain type of event, such as the condition of lactase deficiency—explains the facts under investigation. (Ibid. 305).

Calling abduction "inference to the best explanation" (IBE) obfuscates just what is involved in making such inferences. For instance, as Campos notes, there are at least two species of abductive reasoning: what might be termed "habitual abduction" and "creative abduction" (ibid. 306). And that is just a start—as would be learning how to abduct as the method for diagnosis. Ultimately it requires practice and ongoing learning. What a diagnosis is, is a positive hypothesis (as opposed to a formal hypothesis as in mathematics) and is thus a guess, but not, if one knows what one is doing, a wild or woolly one.

I suggest that learning how to philosophize furthers any journey of discovery. Using Peirce's pragmaticism sheds light on how to abduct, that is, how to create hypotheses which lend themselves to deduction and induction. It gives the steps to follow in identifying the three irreducible categories of First, Second, and Third and of how to distinguish, disassociate, but most importantly, prescind—how, that is, to observe with purpose. It shows how to identify aptness or fit of hypotheses with a purpose for action which in turn suggests how best to proceed reasonably. This step involves engagement with and exploration of concepts that woven from prescinded perceptions leading to the possibility of realization that is the evolving developing universe. Finally, it details how to develop a metaphysical fabric or web that, though it is fallible, is sufficiently stable to support the experimental sciences. Apart from any other consideration, it is an economical method of conducting inquiry that brings clarity to the endeavor, and that assists in avoiding getting bogged down, misled, diverted, or stunted.

Admissions to fallibility require humility but so saying, such open-mindedness is protection against the tyranny of prejudice; the time, money, and life-wasting activity of "beating a dead horse," and any blocks to the freedom of the imagination and creativity. Furthermore, it allows for the process of discovery to move beyond the facts of knowledge to the possibility of information. It is the way of creating or discovering the new. Philosophy, as with all science, is a human endeavor. As with all endeavor it takes time: maybe a couple of thousand years as with disproving Euclid's Fifth Postulate and opening the way to topology; maybe a few moments as with Andy Kirk's prediction of a back draft in time to save his team from fatal disaster; maybe some minutes, hours, days for Denise Levertov to write one of her breathtaking poems; and maybe a lifetime to live ourselves.

Appendix B

The Primary Texts of Charles Sanders Peirce

Grasping Peirce is no mean feat; he is not for the faint-hearted. Other great philosophers are more accessible, but none, past or present, was or is as comprehensive in their method for achieving clarity as Peirce. This is quite ironic: his papers are any editor's nightmare, his style is often cumbersome, and his language is not only littered with neologisms, but he often used different names for the same concept and had a tendency to be inconsistent in their use. In addition, he used many words in ways other than are commonly understood. Yet, as Douglas Anderson wrote in an email to me:

> Clarity can be used in many ways—clear skies, clear vision, clarity of a wine, etc. Peirce focuses on simple logical clarity by which he seems to mean knowing the boundaries of a concept as best we can. He borrows from Descartes and Leibniz who argue that clarity and distinctness are the two modes of measuring a definition. Peirce adds the pragmatic meaning as yet another level of clarity—ultimately for him the final measure of clarity. Thus, the better we know the possible and necessary consequences of "we" the "clearer" will our understanding of the concept be. All of this means for Peirce that there is NO absolute clarity—all concepts have some element of indeterminacy (vagueness or generality). To be intelligible for Peirce means to be something that inquiry might grasp. So Kant's "thing in itself" is unintelligible but "neutrons" are intelligible. Even "God" is intelligible insofar as a belief may have recognizable consequences. But something could be very intelligible and not very clear. And of course no unintelligible "thing" could be clear to us. (Anderson, 2013)

Nominalism and the finious process of causation are the greatest bars to understanding Peirce. To fully grasp him one needs to "get real"—to abandon atemporal approaches. In what is known as his second period, Peirce freely admitted to the error of nominalism in some of his early work and set himself to rectifying this by various means including rewriting whole papers, adding footnotes to some earlier documents, and by way of extending explanation of earlier documents in later documents as a means of improving clarity.

Unfortunately, despite his efforts, Peirce did not manage to make himself clear. Of his papers, numbering some 100,000 pages, around half were undated. This has presented a major problem in understanding his work.

Following his death in 1914, Peirce's philosophical papers were lodged together with his working library with the Harvard University Department of Philosophy by his widow Juliette and the philosopher Josiah Royce. In the decades following their lodgment, a number of attempts were made to bring order to his papers but it was not until 1967 when Richard Robin published his *Annotated Catalogue of the Papers of Charles S. Peirce* followed in 1971 by *The Peirce Papers: A Supplementary Catalogue* that a complete catalog of the Harvard University holdings became available to scholars. Robin began work on this in 1960, two years after the publication of the last volume of the *Collected Papers of Charles Sanders Peirce, Volumes I–VII* edited by Charles Hartshorne and Paul Weiss [1931–1935] and *Volumes VII and VIII* by Arthur Burks [1958]. Unfortunately, Robin did not cross-reference all the material in these eight volumes and no one has, to date, taken on this task.

Reading the preface to Robin's catalog, one can surmise that Hartshorne and Weiss prepared their six volumes from two sources. The first of these was a nine-page typescript "List of C. S. Peirce Manuscripts" by W. F. Kernan who was assisting Josiah Royce in organizing Peirce's papers and collaborating with him on an article entitled "Charles Sanders Peirce" which appeared in the *Journal of Philosophy*, December 21, 1916. Hartshorne and Weiss's other source was a twenty-page typescript, "Notes on Papers and MSs in the Charles S. Peirce Collection" (Lenzen 1917) which is an evaluation of the contents and the physical condition of the manuscripts which at the time were sorted into eighty-three boxes. By the time Burks edited the last two volumes, the Harvard collection had morphed into three separate sets of Peirce materials. The first which made up the bulk of the Peirce Collection consisted of sixty-one boxes and bundles that had been organized, boxed, and cataloged in 1941 by Knight W. McMahan. McMahan's ninety-nine-page typewritten "Catalogue of the C. S. Peirce Manuscripts" was a description of what the boxes contained. The second consisted of some nineteen boxes which had neither been classified nor cataloged. The third set, the correspondence, had been partially organized by McMahan.

A number of smaller collections of his writings were published over the years but the microfilms *The Charles S. Peirce Papers* (1966) and the microfiche *Peirce, C.S. Complete Published Works* (1977) are the only comprehensive source of the Harvard University holdings of his work publicly available.

These are the collections of interest here. No listing of Peirce's library has been made. It is my understanding that following lodgment with Harvard the monographs were put into the stacks. Some have since been identified and the marginalia thereby recovered.

In 1976 the Peirce Edition Project (PEP) was established and has been part of the University of Indianapolis, Purdue's School of Liberal Arts since 1983. Its primary purpose is to organize and date all the manuscripts and to produce an approved scholarly edition of Peirce's writings that can facilitate the study of the historical development of his thought. Over a three-year period, the task of assembling the collections and arranging them chronologically was carried through, though, most unfortunately, the product of this exercise is not available on the public record, as yet. Although PEP's collection, like all the previous ones, is selective, toward the end of each volume a chronological list of all material for the period of the volume in hand is given and an indication is made of what is published in the volume. To date PEP has published seven volumes: *Writings of Charles S. Peirce: A Chronological Edition, vols. 1–6 & 8,* being *W1* (1857–1866), *W2* (1867–1871), *W3* (1872–1878), *W4* (1879–1884), *W5* (1884–1886), *W6* (1886–1890), and *W8* (1890–1892). While it is not their intention to include everything Peirce ever wrote—that would take around 104 volumes—it has been essential to date every fragment of his work. This can only be described as a task of daunting proportions. By example, the Robin Catalogue is sorted into topic areas and, excluding the correspondence, lists around 1,650 "manuscripts" more than half of which are undated. "Manuscript" is, in fact, a loose term for folders and includes notebooks and fragments.

An electronic version of Robin's catalog is now available as is a CD of the *Collected Papers of Charles Sanders Peirce, Volumes I–VII* edited by Charles Hartshorne and Paul Weiss (1931–1935) and *Volumes VII and VIII* by Arthur Burks (1958) (*Peirce: Collected Papers* [1992] Past Masters series, Charlottesville, VA., Intelex). This makes keyword searches possible but does not overcome the myriad of other problems presented by these texts. Although there has been talk of making PEP's *Writings of Charles S. Peirce: A Chronological Edition* in an electronic form, this has not, as yet, eventuated. Nonetheless, even the availability of this work for electronic search would still be less than satisfactory for the very reason that Peirce himself did not achieve a clear single line of thinking. This was exacerbated by his penchant for discursion; for beginning with one idea and making digressions into equally interesting ideas—sometimes not returning to the original idea, but rather picking up again at a later date.

Perhaps what we need, as readers of his work, is an online catalog of his ideas cross-referenced chronologically. For now, though, I have had to make do with what is available.

Glossary

actual (Peirce)

- that which is met within the past, present, and future (Peirce, 1908).

causality

- the relationship between cause and effect.

causation

- the production of an effect by its cause.
- Peirce's theory of causation is most clearly enunciated in his 1902 paper "*On science and natural classes*" (EP2, item 9, 1902).

central limit theorem (Peirce)

- states that random processes, such as rolling dice, the velocity of gas molecules in a closed container, or sampling arbitrarily from a population, will express itself by the Gaussian power law—the familiar bell-shaped curve, or normal distribution as Peirce originally coined the term. Peirce likes to say that this theorem proves "chance begets order" (CP6.297, 1891). More generously interpreted, the central limit theorem suggests that phenomena have a tendency towards self-norming, or as Peirce would put it, "all things have a tendency to take habits. For . . . every conceivable real object, there is a greater probability of acting as on a former like occasion than otherwise" (CP1.409, 1890; see also CP1.390, 1890; CP6.101 1901, CP6.280, 1893). Peirce calls these *finious* processes, in that direction emerges in and through their interactive behavior (see CP7.471, 1908).

critical (Kant)

- the philosophical/rational ability to delimit reason. Kant's first major work, "The Critique of Pure Reason" set out to show the limits of reason so as to open up the possibility of talking meaningfully about morality and religion, matters that do not fall exclusively within the compass of reason. For example, Kant argued that, though reason enables us to think about causation, it does not allow us to think about the cause of causation. Reason will come to contradictory but equally plausible conclusions (antinomies) if it tries to think about things beyond its limits, such as the cause of causation.

excluded middle

- the individual is determinate in regard to every possibility, or quality, either as possessing it or as not possessing it. This is the principle of excluded middle, which does not hold for anything general, because the general is partially indeterminate.

existent

- that interacts with things in a spatiotemporal environment.

fact (Peirce)

- that which the logicians call the *contingent*, that is, the accidentally actual, and whatever involves an unconditional necessity, that is, force without law or reason, *brute* force.

finious (Peirce)

- "tendency toward a final state" (CP7.471, 1898).

habit (Peirce)

- "[Readiness] to act in a certain way under given circumstances and when actuated by a given motive is a habit; and a deliberate, or self-controlled, habit is precisely a belief" (CP5.480, 1907).

hermeneutics (Schleiermacher)

- The branch of knowledge that deals with theories of interpretation, especially of scripture.

heuristic

- In philosophy, especially in Continental European philosophy, the adjective "heuristic" (or the designation "heuristic device") is used when an entity X exists to enable understanding of, or knowledge concerning, some other entity Y. A good example is a model which, as it is never identical with what it models, is a heuristic device to enable understanding of what it models. Stories, metaphors, and so on, can also be termed "heuristic" in that sense. A classic example is the notion of utopia as described in Plato's best-known work, *The Republic*. This means that the "ideal city" as depicted in *The Republic* is not given as something to be pursued, or to present an orientation point for development; rather, it shows how things would have to be connected and how one thing would lead to another (often with highly problematic results), if one would opt for certain principles and carry them through rigorously.
- "Heuristic" is also often commonly used as a noun to describe a rule-of-thumb, procedure, or method philosophers of science have emphasized the importance of heuristics in creative thought and constructing scientific theories.

history of philosophy—approaches

- The history of philosophy can be approached either *exegetically* (in which case the main question is the *interpretive question* of what past philosophers mean and how the structure of their thought holds together) or *critically* (in which case the main question is the *logical question* of whether what past philosophers said was true or false, and what the philosophical consequences of their views are). The word *exegesis* can mean explanation, but as a technical term it means "to draw the meaning out of" a given text. *Exegesis* may be contrasted with *eisegesis*, which means, "to read one's own interpretation into" a given text. In general, *exegesis* presumes an attempt to view the text objectively, while *eisegesis* implies more subjectivity. One may encounter the terms *exegesis* and *hermeneutics* used interchangeably; however, there remains a distinction. An *exegesis* is the interpretation and understanding of a text on the basis of the text itself. A *hermeneutic* is a practical application of a certain method or theory of interpretation, often revolving around the contemporary relevance of the text in question.

hypostatic abstraction, also *hypostasis* or *subjectal abstraction*

- a formal operation that takes an element of information, such as might be expressed in a proposition of the form *X is Y*, and conceives its information to consist in the relation between a subject and another subject, such as expressed in a proposition of the form *X has Y-ness*. The existence of the latter subject, here *Y-ness*, consists solely in the truth of those propositions that have the corresponding concrete term, here *Y*, as the predicate. The object of discussion or thought thus introduced may also be called a *hypostatic object*.

nature

• according to the Enlightenment, by the exercise of reason, one can undo the damage created by religion and society and uncover—discover—the nature that lies waiting to be revealed beneath but obscured by prejudices. For Enlightenment thinkers, nature is the Good, and they assumed that, by stripping away the prejudices of custom, tradition, and religion and replacing those prejudices with the results of the careful use of reason alone, humankind would discover, implicit in nature, *the* aesthetic and ethical goals and standards.

percept

• an object of perception (*M19*).
• the mental product or result of perceiving (as distinguished from action) (*L19*).

precide (Peirce)

• "In those respects in which a sign is not vague, it is said to be definite, and also with a slightly different mode of application, to be precise, a meaning probably due to *præcisus* having been applied to curt denials and refusals. It has been the well-established, ordinary sense of precise since the Plantagenets; and it were much to be desired that this word, with its derivatives *precision, precisive*, and so on, should, in the dialect of philosophy, be restricted to this sense. To express the act of rendering precise (though usually only in reference to numbers, dates, and the like), the French have the verb *préciser*, which, after the analogy of *décider*, should have been *précider*. Would it not be a useful addition to our English terminology of logic, to adopt the verb to precide, to express the general sense, and to render precise? Our older logicians with salutary boldness seem to have created for their service the verb to prescind, the corresponding Latin word meaning only to 'cut off at the end,' while the English word means to suppose without supposing some more or less determinately indicated accompaniment. In geometry, for example, we *prescind* shape from color, which is precisely the same thing as to 'abstract' color from shape, although very many writers employ the verb 'to abstract' so as to make it the equivalent of 'prescind.' But whether it was the invention or the courage of our philosophical ancestors which exhausted itself in the manufacture of the verb 'prescind' the curious fact is that instead of forming from it the noun *prescission*, they took the pattern from the French logicians in putting the word precision to this second use. About the same time (see Watts, *Logick*, 1745, I, vi, 9 ad fin.) the adjective precisive was introduced to signify what prescissive would have more unmistakably conveyed. If we desire to rescue the good ship philosophy for the service of science from the hands of lawless rovers of the sea of literature, we shall do well to keep prescind, presciss, prescission, and prescissive on the one hand, to refer to dissection in hypothesis, while precide, precise, precision, and precisive are used so as to refer exclusively to an expression of determination which is made either full or free for the interpreter. We shall thus do much to relieve the stem 'abstract' from staggering under the double burden of conveying the idea of prescission as well as the unrelated and very important idea of the creation of *ens rationis* out of an [*epos pteroen*]—to filch the phrase to furnish a name for an expression of non-substantive thought—an operation that has been treated as a subject of ridicule—this hypostatic abstraction—but which gives mathematics half its power" (Peirce, 1905b).

premiss (Peirce)

• Peirce used the spelling "premiss" instead of "premise" because "premiss" is derived from the medieval logicians' *praemissa*, while "premise" is properly a legal term.

prescisive abstraction or *prescision* (Peirce)

• variously spelled precisive abstraction or prescission a formal operation that marks, selects, or singles out one feature of a concrete experience to the disregard of others (see *precide* above).

quidditas, entitas, and haecceitas

• *quidditas*
  ◦ the particular form imposed that gives some matter its identity, its *quiddity* or refers to the universal qualities of a thing, its *whatness*, or the aspects of a thing which it may share with other things and by which it may form part of a genus of things.

- *entitas*
  - the term *beingness* translates *entitas*, which is the abstract noun coined to correspond to *ens* ("being") the English cognate ("entity") has a concreteness that is not implied in the Latin term, although it may be, and by Scotus often seemed to be, used in a concrete sense—as one might speak of this white patch as "whiteness."

- *haecceitas*
  - denotes the discrete qualities, properties, or characteristics of a thing which make it a *particular* thing; is the *thisness* of a person or object; whereas haecceity refers to aspects of a thing which make it a *particular* thing, *quiddity*.

reason

- Following Descartes (via Locke), Enlightenment thinkers took reason to be the sense common to human beings. Reason replaced tradition as the "common sense." Reason was assumed to be a matter of method and to be based on indubitable assumptions. Thus, to be rational was to begin with the appropriate premises and then to proceed methodically. In spite of the fact that Enlightenment thinkers took reason to be a natural ability of any human being, most of them also assumed that the average person's reason has been corrupted by the cultural environment and especially by the influence of churches. Churches were considered the most corrupting of influences because churches put revelation above reason and hold that there is something that transcends reason. The enlightened were those who have escaped the thrall of this influence—those who have escaped the, at best, hasty judgments ("prejudices") of religion and everyday culture.

schemata

- in Kantian philosophy, a rule or procedure of the imagination enables the understanding to apply a concept, especially a category, to what is given in sense-perception (*L18*).
- a schematic representation of something, a hypothetical outline or plan; a theoretical construction; a draft; a synopsis; a design (*L19*).
- *psychology:* an (unconscious) organized mental model of something in terms of which new information can be interpreted or an appropriate response made (*E20*).

semiology (Saussure)

- Charles Morris's threefold division of a semeiotics syntactics, semantics, and pragmatics based on a dyadic, positivist reading of Peirce's triadic semeiotic, and a misreading of Peirce's critique of dyadic views of signs and of foundationalism.

semiotics (Morris)

- Peirce's *semeiotic* should not be confused with the semiology of the Swiss linguist, Ferdinand de Saussure whose *Cours de linguistique générale (Course in General Linguistics)* (1916) was published two years after Peirce's death.
- James Liszka points out, "For Saussure, signs are primarily a psychological entity" (1996:15) whereas "Peirce sees semeiotic as leading principles to sciences such as general and social psychology and linguistics" (p. 16). Peirce's semeiotic is normative; Saussure's semiology is empirical, and as Liszka argues (p. 16).
- The only way in which the logical or formal view of semeiotic and the empirical one would be compatible is if the empirical and the formal were treated the same. This is generally called the theory of psychologism; it is something Peirce argues fervently against (see CP2.39-54, c.1902).

significs (Peirce)

- may be defined as the science of meaning or the study of significance, provided sufficient recognition is given to its practical aspect as a method of mind, one which is involved in all forms of mental activity, including that of logic. In Baldwin's *Dictionary of Philosophy and Psychology* [1901–1905] the following definition is given: "1. Significs implies a careful distinction between (a) sense or signification, (b) meaning or intention and (c) significance or ideal worth. It will be seen that the reference of the first is mainly verbal (or rather sensual), of the second volitional, and of the third moral (e.g., we speak of some event 'the significance of which cannot be overrated, and it would be impossible in such a case to substitute the ' sense' or the 'meaning' of such event, without serious loss). Significs treats of the relation of the sign in the widest sense to each of these. 2. A proposed method of mental training aiming at the concentration of intellectual activities on that which is implicitly assumed to constitute the primary and ultimate value of every form of study, i.e., what is at present indifferently called its meaning or sense, its import or significance. . . . Significs as a science would centralize and co-ordinate, interpret, inter-relate and concentrate the efforts to bring out meanings in every form, and in so doing to classify the various applications of the signifying property clearly and distinctly." Since this dictionary was published, however, the subject has undergone further consideration and some development, which necessitate modifications in the definition given. It is clear that stress needs to be laid upon the application of the principles and method involved, not merely, though notably, to language, but to all other types of human function. There is a need to insist on the rectification of mental attitude and increase of interpretative power which must follow on the adoption of the significant view point and method, throughout all stages and forms of mental training, and in the demands and contingencies of life.

stochastic

- randomly determined, that follows some random probability distribution or pattern, so that its behavior may be analyzed statistically but not predicted precisely (*M20*).

tuism (Peirce)

- the doctrine that all thought is addressed to a second person, or to one's future self as to a second person (1891 edition of the *Century Dictionary*).

valency

- might, power, strength (*E-M17*).
- *chemistry:* the power or capacity of an atom or group to combine with or displace other atoms or groups in the formation of compounds, equivalent to the number of hydrogen atoms that it could combine with or displace; a unit of this (*M19*).
- *linguistics:* the power of grammatical elements especially a verb to govern other elements in the same sentence (*L20*).

veracity

- correspondence with truth or fact (*E17*).

verisimilitude

- the appearance of being true or real; likeness or resemblance to truth, or fact; realistic quality; probability (*E17*); a statement; and so on having the mere appearance or show of being true or factual; an apparent truth (*L18*).

verity

- the actuality or reality of something (*M17*).

References

Adriaans, Pieter & Johan van Benthem. (2008) *Philosophy of Information. 'Introduction: Information is What Information Does.'* Charlottesville, VA: University of Virginia. https://www.illc.uva.nl/Research/Publications/Reports/X-2008-10.text.pdf

Alexander, Samuel. (1968) *Beauty and Other Forms of Value.* New York: Thomas Y. Crowell Co.

Almeder, Robert. (1985) 'Peirce's thirteen theories of truth.' *Transactions of the Charles S. Peirce Society* 21(1): 77–94.

Anathaswamy, Anil. (2011) 'Uncertainty entangled.' *New Scientist* 210(2810): 28–31.

Anderson, Douglas R. (1986) 'The evolution of Peirce's concept of abduction.' *Transactions of the Charles S. Peirce Society* 22(2): 145–164.

Anderson, Douglas R. (1987) *Creativity and the Philosophy of C. S. Peirce.* Dordrecht: Nijhoff.

Anderson, Douglas R. (1995) *Strands of System: The Philosophy of Charles Peirce.* West Lafayette, IN: Purdue University Press.

Anderson, Douglas R. (2002) 'Peirce's common sense marriage of religion and science.' SAAP 2002.

Apel, Karl-Otto. (1967, 1981) *Charles S. Peirce: From Pragmatism to Pragmaticism.* Amherst, NY: University of Massachusetts Press.

Aquinas. (2012) *Summa Theologicae.* https://thelycaeum.wordpress.com/2012/11/11/thomas-aquinas-the-fourth-way/

Augustine. (2009) *City of God, Book XI.* http://www.newadvent.org/fathers/1201

Baake, Ken. (2003) *Metaphor and Knowledge: The Challenges of Writing Science.* Albany, NY: State University of NY Press.

Bak, Per & Maya Paczuski. (1995) 'Complexity, contingency, and criticality.' Proc. National Academy of Science USA.

Baldwin, James. (Ed.) (1901–5) *Dictionary of Psychology and Philosophy.* New York: Macmillan.

Bateson, Gregory. (1972) '*Form, Substance and Difference in Steps to an Ecology of Mind.* Chicago, IL: University of Chicago Press, 454–471.

Belenky, Mary Field, Blythe McVicker Clinchy, Nancy Rule Goldberger, et al. (1986) *Women's Ways of Knowing: The Development of Self, Voice and Mind.* New York: Basic Books.

Bentley, Paul. (2001, 2013) 'A matter of integrity: A review of Yuzo Mikami's Utzon's Sphere.' Temporary Works Forum Australia. http://www.twf.org.au/research/mikami.html

Bergson, Henri. (1998 [1911]) *Creative Evolution*, tr., Arthur Mitchell, NY: Dover.

Bernstein, Richard J. (2013) *The New Constellation: The Ethical-Political Horizons of Modernity/Postmodernity.* Maiden, MA: Polity Press.

Binghurst, Robert. (2002) *The Elements of Typological Style.* Vancouver, BC: Hartley & Marks.

Boisvert, Raymond. (2013) 'Camus: Between yes and no.' *Philosophy Now.* Issue 98.

Boler, John F. (1963) *Charles Peirce and Scholastic Realism: A Study of Peirce's Relation to John Duns Scotus.* Seattle, WA: University of Washington Press.

Brill-Edwards, M. (2000) 'Whose health, what protection?' In: M. L. Barer, K. M. McGrail et al. (Eds.), *Tales from the Other Drug Wars.* Vancouver, BC: Centre for Health Services and Policy Research.

Buchanan, Mark. (2000) Ubiquity: *The Science of History...Or Why the World is Simpler than We Think.* London: Weisenfeld & Nicholson.

Bullock, Alan, et al. (Eds.) (1988) *The Fontana Dictionary of Modern Thought.* London: Fontana.

Byers, William. (2007) *How Mathematicians Think: Using Ambiguity, Contradiction and Paradox to Create Mathematics.* Princeton, NJ: Princeton University Press.

Cahill, Reginald. (2005) *Process Physics: From Information Theory to Quantum Space and Matter.* New York: Nova Publisher.

Cahill, Reginald T. (2003) 'Process Physics,' *Process Studies Supplement 2003*, Issue 5, http//www.mountainman.ccom.au/proves_physics/.

Campos, Daniel G. (2009) 'Imagination, concentration and generalization - Peirce on the reasoning abilities of the mathematician.' *Transactions of the Charles S. Peirce Society* 45(2): 135–156.

Casagrande, David. (1999) 'Information as verb: Re-conceptualizing information for cognitive and ecological models.' *Georgia Journal of Ecological Anthropology* 3: 4–13.

Castellani, Brian. (2013) *Map of the Complexity Sciences.* Bloomington, IN: Indiana University. http://scimaps.org/mapdetail/map_of_complexity_sc_154

Chalmers, Alan F. (1982) *What is This Thing Called Science?* 1st edition. Buckingham: Open University Press.

Chalmers, Alan F. (1990) *Science and Its Fabrication.* Buckingham: Open University Press.

Choi, Charles Q. (2017) '*Our Expanding Universe: Age, History & Other Facts.*' New York: Future US Inc. http://www.space.com/52-the-expanding-universe-from-the-big-bang-to-today.html

Christansen, Peder-Voetmann. (1993) 'Peirce as participant in the Bohr-Einstein discussion.' In: Edward C. Moore (Ed.), *Charles S. Peirce and the Philosophy of Science*. Tuscaloosa, AL: University of Alabama.

Colebach, Tim. (2013) "Who got the gains from growth in income?" *Sydney Morning Herald*, October 10, 2013. http://www.smh.com.au/federal-politics/political-news/countrys-rich-have-lions-share-of-income-growth-20131009-2v8q2.html

Cooke, Elizabeth F. (2007) *Peirce's Pragmatic Theory of Inquiry: Fallibilism and Indeterminacy*. New York: Continuum.

Cottingham, John. (Ed.) (2008) *Western Philosophy: An Anthology*. Cambridge, MA: Blackwell.

Cronburg, Turil. (2010) 'The very deep intro to Maslow.' Evoke. April 18.

Crowder, George. (2003) 'Pluralism, relativism and liberalism in Isaiah Berlin.' Australasian Political Studies Association Conference.

Curley, Thomas V. (1969) 'The relation of the normative sciences to Peirce's theory of inquiry.' *Transactions of the Charles S. Peirce Society* 5(2): 90–106.

Davies, Paul. (2007a) 'Taking science on faith.' New York Times: November 24.

Davies, Paul. (2007b) 'The flexi laws of physics.' *New Scientist* 2610: 30 June.

Davies, Paul. (2010) 'Universe from bit.' In: Paul Davies & Niels Henrik Gregersen (Eds.), *Information & the Nature of Reality from Physics to Metaphysics*. Cambridge: Cambridge University Press.

Davies, Paul & Niels Henrik Gregersen. (Eds.) (2010) *Information & the Nature of Reality from Physics to Metaphysics*. Cambridge: Cambridge University Press.

Deacon, Terrence W. (2010) 'The concept of information in biology.' In: Paul Davies & Niels Henrik Gregersen (Eds.), *Information & the Nature of Reality from Physics to Metaphysics*. Cambridge: Cambridge University Press.

de Tienne, André. (2006) 'Peirce's logic of information.' *Seminario del Grupo de Estudios Peirceanos, Universidad de Navarra*.

de Waal, Cornelis. (2005) 'Why metaphysics needs logic and mathematics doesn't: Mathematics, logic and metaphysics in Peirce's classification of the sciences.' *Transactions of the Charles S. Peirce Society* 41(2): 283–297.

DiLeo, Jeffrey R. (1991) 'Peirce's haecceitism.' *Transactions of the Charles S. Peirce Society* 27(1): 79–109.

Donne, John. *Meditation XVII* (1624).

Dorey, E. (2004) 'Conflicts of interest: the close links between public policy and the drug industry.' *Chemistry and Industry*, February 21.

Duff, Michael. (2011) 'Theory of everything: Answering the critics.' *New Scientist*, issue 2815, 1 June.

Dukes, M. N. Graham. (2004) 'W.H.O. and priority medicine: Some notes in the margin.' Medicines for Europe and the World – setting priorities, missing the point. *The Hague*, November 18.

Eagleton, Terry. (2009) *Reason, Faith, and Revolution: Reflections on the God Debate*. New Haven & London: Yale University Press.

Eagleton, Terry. (2011) *Hope Without Optimism*. New Haven & London: Yale University Press.

El-Hani, Niño, João Queiroz & Claus Emmeche. (2009) *Genes, Information and Semiosis.* Tartu, Estonia: Tartu University Press.

Else, Liz. (2010) 'A meadowful of meaning.' *New Scientist*, 21 August, 28–31.

Esposito, Joseph L. (1980) *Evolutionary Metaphysics. The Development of Peirce's Theory of Categories.* Athens, OH: Ohio University Press.

Esposito, Joseph L. (1998) 'Peirce, Charles Sanders Peirce (1839–1914).' In: Paul Bouissac (Ed.), *'Encyclopedia of Semiotics.* Oxford: OUP.

Esposito, Joseph L. (2001) 'Synechism: The keystone of Peirce's metaphysics.' *Digital Encyclopedia of Charles S. Peirce.*

Fann, K. T. (1970) *Peirce's Theory of Abduction.* The Hague: Martinus Nijhoff.

Fisch, Max H. (1986) *Peirce, Semeiotic and Pragmatism.* K. L. Ketner & C. J. Kloesel (Eds.). Bloomington, NY: Indiana University Press.

Forrest, Peter. (1991) 'Aesthetic understanding.' *Philosophy & Phenomenological Research* 51(3): 525–540.

Freeman, Eugene. (Ed.) (1983) *The Relevance of Charles Peirce.* La Salle, IL: Monist Library of Philosophy.

Freese, Katherine. (2014) *The Cosmic Cocktail: Three Parts Dark Matter.* Princeton, NJ: Princeton University Press.

Gefter, Amanda. (2011) 'But what if supersymmetry is wrong?' *New Scientist*, 2804, 19 March.

Gilligan, Carol. (1982) *In a Different Voice: Psychological Theory and Women's Development.* Cambridge, MS: Harvard University Press.

Gorman, Tom. (2003) *The Complete Idiot's Guide to Economics.* London: Penguin.

Gray, John. (1995) *Berlin.* London: Fortana.

Hardy, Henry. (Ed.) (2002) *Liberty: Incorporating Four Essays on Liberty.* Oxford: OUP.

Hausman, Carl R. (1979) 'Value and the Peircean categories.' *Transactions of the Charles S. Peirce Society* 15(3): 203–223.

Hausman, Carl R. (1993) *Charles S. Peirce's Evolutionary Philosophy.* Cambridge: Cambridge University Press.

Hawking, Stephen & Leonard Mlodinow (2010) *The Grand Design.* New York: Bantam Books.

Heidegger, Martin. (1927, 1962) *Being & Time.* Oxford: Basil Blackwell.

Heil, John. (1993) 'Belief.' In: Jonathan Dancy & Ernest Sosa (Eds.), *A Companion to Epistemology.* Oxford: Blackwell.

Heisenberg, Werner. (1958, 1971) *Physics and Philosophy.* Harmondsworth, Middlesex: Penguin.

Hoffmeyer, Jesper. (2010) 'Semeiotic freedom: An emerging force.' In: Paul Davies & Niels Henrik Gregersen (Eds.), *Information & the Nature of Reality from Physics to Metaphysics.* Cambridge: Cambridge University Press.

Hofstadter, Douglas. (1979) *Gödel, Escher, Bach: An Eternal Golden Braid.* London: Penguin.

Hofstede, Geert. (1984) 'The cultural relativity of the quality of life concept.' *The Academy of Management Review* 9(3): 389–398.

Honderich, Ted. (Ed.) (1995) *Oxford Companion to Philosophy*. Oxford: OUP.

Hulswit, Menno. (1997) 'Peirce's teleological approach to natural classes.' *Transactions of the Charles S. Peirce Society* 45(3): 727–772.

Hulswit, Menno. (2002) *From Cause to Causation: A Peircean Perspective*. Dordrecht: Kluwer.

Jantsch, Erich. (1980) *The Self-Organizing Universe: Scientific and Human Implications of the Emerging Paradigm of Evolution*. New York: Pergamon.

Jappy, Tony. (2013) *Introduction to Peircean Visual Semeiotics*. New York: Continuum.

Juarrero, Alicia & Carl A. Rubino. (2008) *'Contents & Introduction' to Exploring Complexity, v.4. Emergence, Complexity and Self-Organization: Precursors and Prototypes*. Goodyear, AZ: ISCE Publishing.

Kager, Patrick & Mark Mozeson. (2000) 'Supply chain: The forgotten factor.' Pharmaceutical Executive USA.

Keller, Evelyn Fox. (1995) *Refiguring Life: Metaphors of 20th Century Biology*. New York: Colombia University Press.

Kenny, Anthony. (2010) *A New History of Western Philosophy*. Oxford: Clarendon Press.

Kent, Beverley E. (1987) *Charles S. Peirce: Logic and the Classification of the Sciences*. Kingston, ON: McGill-Queen's University Press.

Ketner, Kenneth Laine. (2013) *The Published Works of Charles Sanders Peirce*. http://www.pragmaticism.net/works/ bibliography

Kimura, Yasuhiko Genku. (2003) 'Alignment beyond agreement.' *Journal of Integral Thinking for Visionary Action* 1(4): 1–7.

Kitto, Kirsty. (2006) *'Modelling and Generating Complex Emergent Behaviour.'* Adelaide, SA: Flinders University.

Kitto, Kirsty. (2007) *'Quantum Theories as Models of Complexity.'* (Thesis) Adelaide, SA: Flinders University.

Klinger, Christopher M. (2005) *'Process Physics: Bootstrapping Reality from the Limitations of Logic.'* (Thesis) Adelaide, SA: Flinders University.

Kruse, Felicia E. (1991) 'Genuineness & degeneracy in Peirce's categories.' *Transactions of the Charles S. Peirce Society* 27(3): 267–298.

Kupfer, Joseph. (2016) *Guide to the Categorical Imperative*. IA: Iowa State University. https://iastate.app.box.com/v/jkupfer-phil230

Lexchin, J. R. (2005) 'Implications of pharmaceutical industry funding on clinical research.' *Annals of Pharmacotherapy* 39: 194–197.

Liszka, James Jakób. (1996) *A General Introduction to the Semeiotic of Charles Sanders Peirce*. Bloomington, IN: Indiana University Press.

Liszka, James Jakób. (2011) 'Peirce's new rhetoric.' *Transactions of the Charles S. Peirce Society* 36(4): 439–476.

Magee, Bryan. (1997) *Confessions of a Philosopher*. New York: The Modern Library.

Magee, Bryan. (2001) *The Story of Philosophy*. London: DK Publishing.

Marietti, Susanna. (2010) 'Observing signs.' In: Matthew E. Moore (Ed.), *New Essays on Peirce's Mathematical Philosophy*, 147–168.

Mayorga, Rosa. (2007) *From Realism to 'Realicism': The Metaphysics of Charles Sanders Peirce*. Lanham, MD: Rowman & Littlefield.

Merrell, Floyd. (2001) 'Semeiotics versus semiology.' Digital Encyclopedia of Charles S. Peirce. http://www.digitalpeirce.fee.unicamp.br/home.htm (Project supported by FAPESP, grant 97/06018-4.) Hosted on the servers of DCA-FEEC-UNICAMP

Midgley, Mary. (2001) *Science and Poetry*. New York: Routledge.

Mikami, Yuzo. (2001) *Utzon's sphere: Sydney Opera House – How it was Designed and Built*. Tokyo: Shokokusha.

Möllering, Guido. (2001) 'The nature of trust: From Georg Simmel to a theory of expectation, interpretation and suspension.' *Sociology* 35(2): 403–420.

Moore, Edward C. (Ed.) (1993) *Charles S. Peirce and the Philosophy of Science: Papers from the Harvard Sesquicentennial Congress*. Tuscaloosa, AL: University of Alabama Press.

Moore, Matthew E. (Ed.) (2010b) *New Essays on Peirce's Mathematical Philosophy*. Bloomington, IN: Indiana University Press.

Mugglestone, Lynda. (2011) 'A Journey through spin.' Oxford, UK: Oxford University Blog. https://blog.oup.com/2011/09/spin/

New English Bible. (1970) Oxford & Cambridge University Presses.

NIH Office of the Director. (2005) 'NIH announces sweeping ethics reform'. NIH News, February 1, US Department of Health and Human Services.

Nubiola, Jaime. (1995) 'The branching of science according to C. S. Peirce.' 10th International Congress of Logic, Methodology & Philosophy of Science.

O'Connor, J. J. & E. F. Robertson (Last Update February 1996). 'Non-Euclidean Geometry.' MacTutor History of Mathematics: St Andrews, Scotland. https://mathshistory.st-andrews.ac.uk/HistTopics/Non-Euclidean_geometry/

Ogborn, Jon & Edwin F. Taylor. (2005) 'Quantum physics explains Newton's laws of motion.' *Physics Education* 40(1): 26–34.

Paavola, Sami. (2001) *Essential Tensions in Scientific Discovery*. Helsinki: University of Helsinki.

Peat, David. (2008) 'Is there a language problem with quantum physics'. *New Scientist* 2637: 42–43.

Peirce, C. S., & Morris R. Cohen (Ed.). (1923) Chance, love and logic. In *Philosophical Essays*. New York: Harcourt, Brace and World.

Peirce, C. S., Charles Hartshorne, & Paul Weiss (Eds.). *Collected Papers of Charles Sanders Peirce*, vols. 1–6. Cambridge, MA: Harvard University Press.

Volume 1: *Principles of Philosophy* (1931).

Volume 2: *Elements of Logic* (1932).

Volume 3: *Exact Logic* – published papers (1933).

Volume 4: *The Simplest Mathematics* (1933).

Volume 5: *Pragmatism and Pragmaticism* (1934).

Volume 6: *Scientific Metaphysics* (1935).

Peirce, C. S., & Arthur W. Burks (Ed.). *Collected Papers of Charles Sanders Peirce*, vols. 7–8. Cambridge, MA: Harvard University Press.

Volume 7: *Science and Philosophy* (1958).

Volume 8: *Reviews, Correspondence and Bibliography* (1958).

Peirce, C. S. Buchler, Justus (Ed.) (1940, 1955) *Philosophical Writings of Peirce*. New York: Dover.

Peirce, C. S. Wiener, Philip P. (Ed.) (1958) *Charles S. Peirce Selected Writings: Values in a Universe of Chance,* NY: Dover,

Peirce, C. S. (1966) *The Charles S. Peirce Papers* in Houghton Library.

Peirce, C. S. Harwick, Charles S. (Ed.) (1976) *Semiotic and Significs: The Correspondence between Charles S. Peirce and Victoria Lady Welby.* Bloomington, IN: Indiana University Press.

Peirce, C. S. Peirce Edition Project. (Eds.) *Writings of Charles S. Peirce: A Chronological Edition,* vols. 1–6 & 8. Bloomington, IN: Indiana University Press.

Volume 1: *1857-1866.* Max H. FISCH et al. (Eds.) (1982)

Volume 2: *1867-1871.* Edward C. MOORE et al. (Eds.) (1984)

Volume 3: *1872-1878.* Christian J. W. KLOESEL et al. (Eds.) (1986)

Volume 4: *1879-1884.* Christian J. W. KLOESEL et al. (Eds.) (1989)

Volume 5: *1884-1886.* Christian J. W. KLOESEL et al. (Eds.) (1993)

Volume 6: *1886-1890.* Nathan HOUSER et al. (Eds.) (2000)

Volume 8: *1890-1892.* Nathan HOUSER et al. (Eds.) (2010)

Peirce, C. S. Ketner, Kenneth Laine. (Ed.) (1992) *Reasoning and the Logic of Things: The Cambridge Conferences Lectures of 1898.* Cambridge, MA: Harvard University Press.

Peirce, C. S. Houser, Nathan et al. (Eds.) *The Essential Peirce. Selected Philosophical Writings,* vols. 1–2. Bloomington, IN: Indiana University Press.

Volume 1: *1867-1892.* Nathan Houser & Christian J. W. Kloesel. (Eds.) (1992)

Volume 2: *1893-1913.* Nathan Houser et al. (Eds.) (1998)

Peirce, C. S. Turrisi, Patricia Ann. (Ed.) (1997) *Pragmatism as a Principle and Method of Right Thinking: The 1903 Harvard Lectures on Pragmatism.* New York: State University of NY.

Peirce, C. S. Moore, Matthew E. (Ed.) (2010a) *Philosophy of Mathematics: Selected Writings Charles S. Peirce.* Bloomington, IN: Indiana University Press.

Peirce, C. S. Bergman, Mats, & Sami Paavola. (Eds.) (2014) *The Commens Dictionary of Peirce's Terms.* Helsinki: Helsinki University Press. http://www.commens.org/dictionary

Peirce, C. S. (1868) 'Some consequences of four incapacities.' *Journal of Speculative Philosophy* 2: 140–157.

Peirce, C. S. (1877) 'Fixation of Belief.' *Popular Science Monthly* 12: 1–15.

Peirce, C. S. (1878) 'How to make our ideas clear.' *Popular Science Monthly* 12: 286–302.

Peirce, C. S. (1891) 'The architecture of theories.' *The Monist* 1: 161–176.

Peirce, C. S. (1893) 'Evolutionary love.' *The Monist* 3: 176–200.

Peirce, C. S. (1905a) 'What pragmatism is.' *The Monist* 15: 161–181.

Peirce, C. S. (1905b) 'Issues of pragmaticism.' *The Monist* 15: 481–499.

Peirce, C. S. (1908) 'A neglected argument for the reality of God.' *The Hilbert Journal* 7: 90–112.

Perlmutter, Saul. (2001) 'Supernovae, dark energy, and the accelerating universe: The status of the cosmological Parameters.' Stanford, CA: Stanford University. https://www.slac.stanford.edu/econf/C990809/docs/perlmutter.pdf

Pfeffer, Jeffrey. (1998) 'Seven practices of successful organizations.' *California Management Review* 40(2): 96–124.

Pickering, J. (2016) 'Is nature habit-forming?' In: Donna E. West & Myrdene Anderson (Eds.), *Consensus on Peirce's concept of habit: Before and Beyond Consciousness*. (Studies in Applied Philosophy, Epistemology and Rational Ethics [SAPERE].) New York: Springer.

Piercy, Marge. 'In the men's room(s).' In: Marge Piercy (Ed.), *Eight Chambers of the Heart: Selected Poems*. Harmondsworth, Middlesex: Penguin Books.

Polzin, Jackie. (2021) *Brood* London: Picador.

Potter, Vincent G. (1967, 1997) *Charles S. Peirce on Norms and Ideals*. [First published in 1967] New York: Fordham University Press.

Raposa, Michael L. (1989) *Peirce's Philosophy of Religion*. Bloomington, IN: Indiana University Press.

Raz, Joseph. (2003) *The Practice of Value*. Oxford: Clarendon Press.

Reimer, Bennett. (1970) *A Philosophy of Music Education*. Englewood Cliffs, NJ: Prentice-Hall.

Rescher, Nicholas (1996) *Process Metaphysics: An Introduction to Process Philosophy*. State University of NY.

Reynolds, Andrew. (2002) *Peirce's Scientific Metaphysics: The Philosophy of Chance, Law and Evolution*. Nashville, TN: Vanderbilt University Press.

Robin, Richard D. (1967 [supplement 1971]) *Annotated Catalogue of the Papers of Charles S. Peirce*. http://www.iupui.edu/~peirce/robin/robin.htm catalogue.

Rosensohn, William L. 1974. *The Phenomenology of Charles S. Peirce*. Amsterdam: B. R. Grüner.

Royce, Josiah with W. Fergus Kenan. (1916) 'Charles Sanders Peirce.' *Journal of Philosophy, Psychology and Scientific Methods* 13: 701–709.

Sadrā, Mullā. (1992) *The Metaphysics of Mullā Sadrā. Kitab al-masha'ir (The Book of Metaphysical Penetrations)*. New York: The Society for the Study of Islamic Philosophy & Science.

Sartre, Jean-Paul (1958 [1943]) *Being and Nothingness: An Essay on Phenomenological Ontology*. London: Methuen.

Saussure, Ferdinand de. (1916) *Cours de linguistique générale. (Course in General Linguistics.)* Paris: Payot.

Schiller, Friedrich. (1801, 1954) *On the Aesthetic Education of Man in a Series of Letters*. New York: Frederick Ungar Publishing.

Searle, John. (2003: 1) 'Contemporary philosophy in the United States'. In: Nicholas Bunning & E. P. Tsui-James (Eds.), *The Blackwell Companion to Philosophy*. Oxford: Blackwell.

Shorter Oxford English Dictionary. (2002) *Shorter Oxford English Dictionary* [2 vols.]. Oxford: OUP.

Simmel, Georg. (1900, 1990) *The Philosophy of Money*. London: Routledge.

Skagestad, Peter. (1981) *The Road of Inquiry. Peirce's Pragmatic Realism*. New York: Columbia University Press.

Slezak, Michael. (2012) 'Higgs boson is too saintly and supersymmetry too shy.' *New Scientist* 2892: 2.

Smith, Andrew. (2009) 'Truth, negation and the limit of inquiry.' SAAP 2010 Conference.

Smith, John E. (1952) 'Religion and theology in Peirce'. In: Philip P. Weiner* & Frederick H. Young (Eds.), *Studies in the Philosophy of Charles Sanders Peirce: First series*, 251–270.

Smith, John E. (1978) *Purpose and Thought. The Meaning of Pragmatism.* Chicago, IL: University of Chicago Press.

Smith, John E. (1980) 'Community and reality.' In: Eugene Freeman (Ed.), *The Relevance of Charles Peirce*, 38–58.

Smyth, Richard A. (2002) 'Peirce's normative science revisited.' *Transactions of the Charles S. Peirce Society* 38(1&2): 283–306.

Sorrell, Kory S. (2004) *Representative Practices: Peirce, Pragmatism and Feminist Epistemology.* New York: Fordham University Press.

Spiegelberg, Herbert. (1956) 'Husserl's and Peirce's phenomenologies.' *Philosophy & Phenomenological Research* 17(2): 164–185.

Spiers, Hugo, & Jenny Gimple. (2003) *The Human Mind.* Open University, as presented by Robert Winston & produced by BBC1.

Stanley, Donald E., & Daniel G. Campos. (2013) 'The logic of medical diagnosis.' *Perspectives in Biology and Medicine* 56(2): 300–315.

StGeorge, Michael. (2002) 'Survival of a fitting quotation.' http://www.silkworth.net /fitquotation.pdf

Stiglitz, Joseph E. (2013) *The Price of Inequality.* London: Penguin.

Suber, Peter. (1997) 'Symbolic Logic: course notes.' Dept. of Philosophy, Earlham College, Richmond, IN.

Sydney Morning Herald. (1986) 9 December.

Szymborska, Wisława. (1990) 'π' (Translation from Polish). In: Adam Czerniawski (Ed.), *People on a Bridge: Poetry of Wisława Szymborska.* London: Forrest Books.

Tiercelin, Claudine. (2010) 'Peirce on mathematical objects and mathematical objectivity.' In: Matthew E. Moore (Ed.), *New Essays on Peirce's Mathematical Philosophy*, 81–122.

Tillich, Paul. (1957) *Systematic Theology: Existence and the Christ.* Chicago, IL: University of Chicago Press.

Thagard, Paul. (1990) 'Philosophy and machine learning.' *Canadian Journal of Philosophy* 20(2): 261–276.

University of New Hampshire, Office of Intellectual Property Management. (2002) 'Bayh-Dole Act.' www.unh.edu/oipm

Wang, Hao. (1990) *Reflections on Kurt Gödel.* Cambridge, MA: MIT Press.

Watts, Isaac. (1745) *Logick.* London: Longman, Shewell & Blackstone.

Whitney, William Dwight. (Ed.) (1889–1891) *Century Dictionary*, first edition. New York: The Century Company of NY.

Wilczek, Frank. (2008) *The Lightness of Being: Mass, Ether and the Unification of Forces.* New York: Basic Books.

Williams, Bernard. (1972) *Morality: An Introduction to Ethics.* Cambridge: Cambridge University Press.

Wright, Judith. (1971, 1994) 'Eve to her daughters.' In: Judith Wright (Ed.), *Collected Poems 1942–1985.* Sydney: Angus & Robertson.

Yu, Chong Ho. (1994) 'Abduction deduction induction.' Annual Meeting of American Educational Research Association.

Name Index

Subject Index

About the Author

Dorothea Sophia has a BEd (primary), an MBA, and a PhD (Philosophy). She has spent most of her working life in administration in small enterprises, large corporations, unions, various media organizations, and the university sector. She lives in Sydney.

Ingram Content Group UK Ltd.
Milton Keynes UK
UKHW011450160323
418682UK00006B/92

9 781793 654106